Creating a Culture of Accessibility in the Sciences

Creating a Culture of Accessibility in the Sciences

Mahadeo A. Sukhai
University of Toronto and
Ontario Cancer Institute,
Toronto, ON, Canada

and

Chelsea E. Mohler
National Educational Association of
Disabled Students (NEADS),
Ottawa

AMSTERDAM • BOSTON • HEIDELBERG • LONDON
NEW YORK • OXFORD • PARIS • SAN DIEGO
SAN FRANCISCO • SINGAPORE • SYDNEY • TOKYO
Academic Press is an imprint of Elsevier

Academic Press is an imprint of Elsevier
125 London Wall, London EC2Y 5AS, United Kingdom
525 B Street, Suite 1800, San Diego, CA 92101-4495, United States
50 Hampshire Street, 5th Floor, Cambridge, MA 02139, United States
The Boulevard, Langford Lane, Kidlington, Oxford OX5 1GB, United Kingdom

Notices
Knowledge and best practice in this field are constantly changing. As new research and experience
broaden our understanding, changes in research methods, professional practices, or medical
treatment may become necessary.

Practitioners and researchers must always rely on their own experience and knowledge in
evaluating and using any information, methods, compounds, or experiments described herein. In
using such information or methods they should be mindful of their own safety and the safety of
others, including parties for whom they have a professional responsibility.

To the fullest extent of the law, neither the Publisher nor the authors, contributors, or editors,
assume any liability for any injury and/or damage to persons or property as a matter of products
liability, negligence or otherwise, or from any use or operation of any methods, products,
instructions, or ideas contained in the material herein.

British Library Cataloguing-in-Publication Data
A catalogue record for this book is available from the British Library

Library of Congress Cataloging-in-Publication Data
A catalog record for this book is available from the Library of Congress

ISBN: 978-0-12-804037-9

For Information on all Academic Press publications
visit our website at https://www.elsevier.com

Working together
to grow libraries in
developing countries

www.elsevier.com • www.bookaid.org

Publisher: Sara Tenney
Acquisition Editor: Mary Preap
Editorial Project Manager: Mary Preap
Production Project Manager: Chris Wortley
Designer: Matthew Limbert

Typeset by MPS Limited, Chennai, India

Dedicated to those who will come after us: Young people with disabilities everywhere, who are interested in and passionate about science. Follow your dreams, and through perseverance, make them a reality.

Contents

Epigraph		xvii
List of Contributors		xix
About the Authors		xxi
Acknowledgments		xxiii
Introduction		xxv

Part I   Students with Disabilities in the Sciences   **1**

1   The landscape for students with disabilities in the sciences   3
Introduction   4
Some key terms   5
Exclusive, segregated, integrated, and inclusive education   6
Differences between undergraduate and graduate laboratory environments   7
Work in the Academic Environment   8
Participation of students with disabilities in the sciences   8
Glass ceilings in the STEM training pipeline   10
What is a "culture of accessibility?"   11
Definition of practical space environments   12
Application of best practices across disciplines   12
Previous forays in accessibility and STEM   13
Conclusion   14

2   Accessibility and science, technology, engineering, and mathematics—the global perspective   15
Introduction   16
Accessibility of science, technology, engineering, and mathematics (STEM) education: a human-rights framework   17
United National convention on the rights of persons with disabilites   18
Higher education in the context of the global landscape of disability rights legislation   19
Cultural perspectives about disability   20
Science as an international endeavor   21
Conclusion   22

Part II   Barriers Faced by Students with Disabilities in the Sciences   **23**

3   Barriers faced by students with disabilities in science laboratory and practical space settings   **25**
   Introduction   27
   Occupational choice   28
   Lack of professional development for service providers and educators   28
      Educators   28
      Service providers   29
   Structural differences in student support systems between high school
      and postsecondary education   29
   Student awareness of support systems in postsecondary   30
   Student engagement with support systems in the educational setting   30
   Lack of access to assistive technologies   31
   Adapting mainstream technologies   31
   Universal access to scientific materials   32
   Availability of accessible formats   32
   Logistical considerations in the lab and classroom   33
      Logistical considerations in the STEM laboratory, fieldwork and
         practicum environments   33
      Logistical considerations in the STEM classroom   34
   Self-advocacy   34
   Support network advocacy   35
   Attitudes   35
   Competing priorities in education   36
   The "gatekeeper function"   37
      The challenge of misinformation   37
      The challenge of lack of information   38
      The challenge of inductive reasoning   38
   Conclusion   38

4   Student perspectives on disability—impact on education, career path, and accommodation   **41**
   Introduction   41
   "What could they do for me?"   42
   "Ignorance is bliss"   43
   "Raising the bar"   45
   "There is always a Way, it's just a matter of finding it"   47
   "Easier said than done"   48
   "Did my opinion matter?"   49
   "Not being good enough"   49
   "Knowledge is preparedness"   52
   "Disability is not something to be ashamed of"   53

"What could have been?"   53
Conclusion   54

5   **Key role of education providers in communication with students
     and service providers**   57
     Introduction   59
     First contact: how faculty find out students with disabilities are in
        their courses   60
     Faculty responsibilities around communication and content delivery   60
     Scope of the teaching team   62
     Some typical situations   62
        Case study #1: undergraduate lab course-based (with pitfalls)   62
        Case study #2: undergraduate lab course-based (ideal scenario)   63
        Case study #3: graduate research environment (with pitfalls)   64
        Case study #4: graduate research environment (ideal scenario)   65
     Conclusion   66

Part III   Student as Educator   69

6   **Disclosure in the sciences**   71
     Introduction   73
     Definition of disclosure   73
     Disclosure of accommodation need vs. disclosure of disability   74
     Process of disclosure   75
     The choice to disclose   75
     Self-advocacy and disclosure   76
     What students should know prior to disclosure   77
     Timing of disclosure   77
     A rubric for disclosure   78
     Identifying the right players   78
     Impact of disability types on disclosure   79
     Pros and cons of disclosure   79
     Implications of disclosure in laboratory and practical space environments   80
     Other factors affecting disclosure   80
     Conclusion   81

7   **Student as ACTor—recognizing the importance of advocacy,
     communication, and trailblazing to student success in STEM**   83
     Introduction   85
     Student as advocate   87
     Student as communicator   88
     Student as trailblazer   89
     Steps for a successful secondary school to postsecondary transition   90
     Conclusion   91

8   **Mental health and well-being for students with disabilities**
     in the sciences   **93**
     Introduction   **95**
     The stress of being a trailblazer   **96**
     Impostor syndrome   **97**
     Stress of disclosure, accommodation, and disability management   **99**
     Disability and the stresses of STEM training   **100**
     Recognizing signs of stress   **101**
     Strategies to improve well-being and maintain balance   **102**
     Conclusion   **103**

9   **Peer-support networks**   **105**
     Introduction   **106**
     Why is peer support important?   **106**
     What is peer support?   **107**
     Types of peer support   **108**
     Benefits of having a peer-support network   **109**
     Potential challenges in having or maintaining a peer-support network   **109**
     Operational issues: ensuring effective peer support   **110**
     Peer support versus peer mentorship   **110**
     Peer support: the beginning of the conversation   **111**
     Where do we find our peers?   **111**
     Virtual (long-distance) peer support   **112**
     Peer support outside disability   **112**
     Conclusion   **113**

Part IV   Student as Learner   **115**

10   **Essential requirements and academic accommodations**
     in the sciences   **117**
     Introduction: what is an essential requirement?   **119**
     Measurement of essential requirements   **120**
     Importance of essential requirements   **121**
     Essential requirements and the evolution of the sciences   **121**
     Relationship between essential requirements and accommodation   **122**
     Conversations around essential requirements and accommodations   **123**
     Reasonable accommodation and undue hardship   **123**
     The "gatekeeper function"   **124**
     Mythbusting accommodations and essential requirements   **125**
       Myth: Accommodations are expensive   **125**
       Myth: Accommodations take away from understanding of course
         or program content   **125**
       Myth: Accommodations are synonymous with "having someone
         else do the work for you"   **125**

Myth: Accommodations are unfeasible to implement for just
  one person   126
Myth: Accommodations do not mirror the "real world"   126
Myth: Accommodations will compromise safety   126
Myth: Accommodations will interfere with timeliness   126
Essential requirements in STEM environments   127
Conclusion   128

11 Universal design for learning   **129**
Introduction   130
Barriers to learning   131
Principles of universal design for learning   132
Applying UDL to the learning environment   133
The role of technology in implementing UDL   134
UDL and the need for accommodation   135
Conclusion   136

12 Inclusive teaching practices   **139**
Introduction   140
Principles of inclusion   140
Identifying implicit expectations and making them implicit   142
Approaches to developing inclusive teaching practices   144
Principles of effective accommodation   145
  Accommodations are individualized   145
  Individualized accommodations are flexible   145
  Accommodations must be developed and implemented in a
    timely manner   146
Conclusion   148

Part V   Students as Mentees, Trainees, and Leaders   **149**

13 Faculty mentorship of students with disabilities in the sciences   **151**
Introduction   153
The importance of mentorship in STEM   154
What is mentorship?   155
Mentors can be people with or without disabilities   156
Forms of mentorship   156
  In-person (one-on-one) mentorship   157
  Online mentorship   157
  Senior/faculty mentorship   157
  Peer mentorship   158
Qualities of a good mentor   158
  A good mentor is proactive   158
  A good mentor is responsive   159

A good mentor is open-minded   159
A good mentor is creative   159
Benefits of becoming a mentor   160
Selecting a good mentor   160
Conclusion   161

14   Faculty supervision of students with disabilities in the sciences   163
Introduction   165
The foundation of the student–supervisor relationship   165
Disability, disclosure, and the student–supervisor relationship   166
Quality of the student–supervisor relationship   167
Deterioration in the student–supervisor relationship   168
Clarifying expectations in the student–supervisor relationship   168
Students in crisis   169
Supervisor's knowledge of and/or willingness to participate in
    disability-related processes   170
The role of the supervisor's knowledge of the interface between
    essential requirements and academic accommodations   171
At the interface of research integrity and accommodations:
    authorship issues   171
Supervisors may assist students with academic and social integration   172
Boundary issues   173
Funding issues   173
Delegated supervision   174
Conclusion   174

15   The student in a leadership, mentorship, and supervision role   175
Introduction   177
Disability management and nontraditional learning environments   178
Employment for students in STEM graduate programs   178
Widening the circle: disclosure when the student is not in a
    traditional learning environment   179
Identifying accommodation needs in nontraditional learning
    environments   180
Achievement of necessary competencies   182
Stress and nontraditional learning environments   183
Student as trailblazer   183
The student's lived experience with disability—impact on perspectives   184
Conclusion   185

16   Leveraging professional development and networking opportunities   187
Introduction   190
Types of networking opportunities   190
Disability-specific structured professional networking activities   191
Networking through conferences, career events, and symposia   192
Peer networks and collaborations   192

Informational interviews   193
Creating your own portfolio of networking opportunities   193
Framing disability in networking   194
Defining your personal story or brand and its impact on networking   194
Receptivity of your network   195
Conclusion   196

Part VI   Accommodating Students With Disabilities in the Sciences   197

17   Accommodating students with disabilities in science laboratories and in fieldwork   199
Introduction   200
Teaching practices, supports, and accommodations   202
Activities in a teaching lab setting in the sciences   202
Accommodation in the graduate research laboratory   203
Accommodation in the fieldwork setting   204
Discussing accommodations in the science lab or fieldwork settings   205
Conclusion   205

18   Human accommodation—laboratory/technical assistants in the sciences   207
Introduction   209
The essential requirements argument   210
Does technical assistance provide an "unfair advantage?"   212
Does technical assistance provide the student "too much help?"   212
Is Technical assistance "unrealistic?"   213
Does technical assistance mean the credit belongs to the assistant?   214
Is technical assistance too expensive?   215
Does the student gain the appropriate learning from having a technical assistant?   216
When to utilize technical assistance?   216
Finding an appropriate technical assistant   217
    Defining the role   217
    Job posting   218
    Interviews   218
    Training   218
    Evaluation   219
Conclusion   219

19   Mainstream technology as accessibility solutions in the science lab   221
Introduction   223
Case Study #1: Using robotics to handle small volumes in a biological sciences research lab   224

Case Study #2: Microscope slide scanners in anatomy, histology,
   or physiology teaching labs   224
Case Study #3: Universal design in the lab—computer-aided
   instrumentation in a physics laboratory   225
Case Study #4: Adaptive technology brought mainstream   225
Case Study #5: Mainstream technology adapted for accessibility 226
Case Study #6: Accessible mainstream technology solutions—build with
   universal design principles in mind   226
Case Study #7: Low-tech solutions   226
Determining the best technology solution for a student with a disability 227
Student considerations   228
Engineering custom technology solutions: the need for a knowledge base 228
Conclusion   229

20 **Assistive technology**   **231**
Introduction   232
What is assistive technology?   233
Barriers to accessing assistive technology   234
Assistive technology in the laboratory setting   235
Meeting the challenges   236
Conclusion   236

21 **Accessible formats in science and technology disciplines** **239**
Introduction   240
What are print disabilities?   240
What are accessible formats?   241
Accessible format materials and technologies   241
Challenges with accessing accessible formats in the classroom setting 242
Accessing accessible format materials in the laboratory setting 243
Tips for making accessible formats accessible in the classroom
   and laboratory   244
Accessibility and online learning environments   245
The Marrakesh Treaty   246
Conclusion   247

22 **Simulation learning**   **249**
Introduction   250
Virtual learning in the sciences   251
Simulations and learning styles   252
Considerations when applying simulation learning to the sciences 252
Simulation learning and accessibility   253
Simulation learning as accommodation   253
Simulation learning as a course/program component   254
Simulation learning in postsecondary education   254
Conclusion   255

23 Physical access in science laboratories   **257**
Introduction   **258**
Considerations for physical accessibility in science labs   **259**
Universal design and physical accessibility of science laboratories   **259**
Best practices   **260**
Accessibility and safety   **261**
Conclusion   **262**

Part VII   Synthesis   263

24 Practicum placements   **265**
Introduction: overview of practicums   **266**
Review of relevant legislation and duties to accommodation   **267**
Introduction to the practicum accommodations process model   **268**
Contextual considerations   **270**
    National context   **270**
    University context   **270**
    Practicum site context   **271**
    Student context   **272**
Partnerships required for successful practicum placement learning
    for students with accommodations   **272**
    Student   **273**
    University faculty   **273**
    University services   **273**
    Practicum site   **274**
Practicum accommodations process   **274**
    Initial meeting of student and course instructor to review
      expectations and explore options   **274**
    Accommodations assessment/recommendations   **276**
    Explore potential practicum partners   **276**
    Finalize specific practicum arrangements   **277**
    Monitor performance throughout practicum   **277**
    Reflect on teaching and learning   **278**
Considerations and strategies for successful provision of practicum
    accommodations   **278**
    Maintenance of academic and professional standards **279**
    Time   **280**
    Equitable opportunities for all students   **281**
Student accommodations   **281**
    Altered time   **281**
    Geographical location   **281**
    Provision of one primary preceptor   **282**
    Homogenous practice area   **282**
    Use of technology   **282**

Considerations for national and international practicums   282
Conclusion   283

**25 General principles of designing accessible learning environments
     in the sciences   285**
Introduction   286
Practical spaces revisited   287
Differences between practical spaces and traditional science laboratories   287
Case study: Practical spaces in occupational and physical therapy   287
Case study: Archival spaces   289
The diversity of practical spaces in STEM education   289
Guiding principles for designing accessible learning environments
   in STEM   290
Overview of universal design principles   291
Flexibility   292
Dynamism   293
Collaboration   293
Fostering positive relationships   294
Does not contravene academic or professional rigor   294
Encompasses the many faces of a student in STEM   295
Conclusion   295

Conclusion: STEM and disability—a vision for the future   297
Bibliography   301
Index   309

Epigraph

Success is a collaboration

Dr. Suzanne Kamel-Reid

List of Contributors

Jeremiah Bach National Educational Association of Disabled Students, Toronto, ON, Canada *(Chapter 9)*

Donna Barker University of Toronto, Toronto, ON, Canada *(Chapter 24)*

Diane Bergeron Canadian National Institute for the Blind, Ottawa, ON, Canada *(Chapter 2)*

Emily M. Duffett Acadia University, Halifax, NS, Canada *(Chapter 6)*

Tina Doyle University of Toronto at Scarborough, Toronto, ON, Canada *(Chapters 17, 23)*

Garrett Everding WeMesh Inc., Kitchener, ON, Canada *(Chapters 19, 22)*

Penny Hartin World Blind Union, Toronto, ON, Canada *(Chapter 2)*

Ainsley R. Latour Memorial University of Newfoundland, Whitby, ON, Canada *(Chapters 12, 14)*

Anuya Pai University of Calgary, Calgary, AB, Canada *(Chapter 25)*

Amanda Philpot New England College, Henniker, NH, United States *(Chapters 2, 4)*

Dayna Schnell University of Calgary, Calgary, AB, Canada *(Chapters 7, 9)*

Amy E. Soden YMCA of Greater Toronto, Toronto, ON, Canada *(Chapter 16)*

Jill Stier University of Toronto, Toronto, ON, Canada *(Chapter 24)*

Jose Viera World Blind Union, Toronto, ON, Canada *(Chapter 2)*

About the Authors

Dr. Mahadeo Sukhai is the world's first congenitally blind cancer geneticist. Dr. Sukhai is currently a team leader with the Advanced Molecular Diagnostics Laboratory at the University Health Network in Toronto, Canada, where he leads the lab's Variant Interpretation Group. In this role, he is responsible for the development and implementation of new methods to interpret the results of genomic tests for cancer, based on the results of the worldwide efforts to understand the cancer genome. As part of this effort, he is one of a team of scientists at the forefront of bringing new genomic technologies into the clinical setting.

Prior to assuming this role, Dr. Sukhai completed his PhD in cancer biology from the University of Toronto (2007), and two postdoctoral fellowships, in genomics and drug discovery, at the University Health Network. Dr. Sukhai also has a significant interest in higher education and the nature of the research training environment, with an emphasis on students and research trainees with disabilities.

Outside of a distinguished research and teaching career, Dr. Sukhai places a strong emphasis on voluntarism, science education, and mentorship. He has been an active volunteer with the International Association of Lions Clubs (1993–2003), the Let's Talk Science Partnership Program (2007–12), the University of Toronto (2001–present), the Canadian National Institute for the Blind (CNIB; 2007–present), and the National Educational Association of Disabled Students (NEADS; 2004–present).

Dr. Sukhai has held numerous volunteer executive leadership positions at local, regional, national, and international levels, and currently serves as senior advisor to the NEADS Board of Directors. Dr. Sukhai recently completed a 3-year term as vice-chair of the Board of Directors of the National Postdoctoral Association (NPA). Dr. Sukhai joined the Research Committee of the CNIB in 2009, the National Board of Directors in 2012, and is currently the chair of the CNIB's Public Education and Advocacy Committee as well as a mentor to CNIB's National Youth Council.

Dr. Sukhai is also the chair of the National Taskforce on the Experience of Graduate Students with Disabilities, established by NEADS, where he led a nationwide intensive multistakeholder effort to understand the barriers faced by graduate students with disabilities in Canadian postsecondary education—a landmark, first-of-its-kind study with significant national and international impact. He is also the executive director of the NEADS National Student Awards Program, Canada's only nationwide cross-discipline and cross-disability scholarship program, and is the principal investigator of a series of projects exploring the culture of accessibility in science laboratories and other practical spaces.

Dr. Sukhai was most recently a member of the 2015 Governor General's Canadian Leadership Conference, and has been recognized numerous times for his contributions to science, higher education, and volunteerism.

Chelsea Mohler is a researcher, educator, advocate, and passionate scholar who happens to be legally blind. She holds a Master's of Science in Occupational Sciences from Western University, Canada. Her graduate thesis work focused on the process persons with vision loss implement to maintain gainful employment. Chelsea's research takes an occupational justice lens, i.e., she uses constructs of social justice and social inequality to understand the equitable opportunity and resources that enable people's engagement in meaningful occupations. This framework views participation in occupation as a human right. Her use of this critical, rights-based framework is informed by her lived experience navigating higher education and employment as a person with a disability.

During her time at Western, Chelsea was chair of the graduate equity committee and commissionaire for students with disabilities. In her role in these positions, she worked to advocate on behalf of graduate students with disabilities. Chelsea has previously held the position of research assistant at The Canadian National Institute for the Blind (CNIB), and currently works for the National Educational Association of Disabled Students (NEADS). In her role with NEADS, Chelsea is a research consultant on two projects. One focuses on the transition into postsecondary for students with disabilities, and the other focuses on a student mentorship model aimed at educating employers on creating a culture of accessibility and inclusion in the workplace. Additionally, Chelsea has coauthored articles exploring the culture of inclusive education for students with visible and invisible disabilities, in the context of the graduate education environment.

Outside of her research interests, Chelsea places a strong emphasis on volunteering, leadership, engagement, and mentorship for youth with disabilities. Chelsea is an active volunteer with the CNIB where she hosts workshops for youth with vision loss focusing on the importance of engaging in community volunteer work throughout secondary and postsecondary. Chelsea is also an active peer mentor with the CNIB where she educates those new to living with vision loss around adapting to their disability. Chelsea is also a tutor and mentor at University of Toronto's Disability Office. In her capacity as mentor, she aims to empower students to become engaged advocates and leaders within their community.

Through her research and scholarship, Chelsea strives to educate students and professionals about inclusion and diversity; her aim is to empower educators and students to have a more inclusive mindset around education, using her own personal experience as a tool to increase awareness. Chelsea's passion for inclusive and accessible educations stems from her own journey as a researcher and academic living with vision loss. Chelsea believes that through telling her own story, and sharing her own experiences, she can encourage students and educators to make positive change.

Acknowledgments

The authors wish to thank all the contributors who have assisted with the manuscript for this book, and have brought their various skillsets, backgrounds, and areas of expertise to this work in a variety of ways. We also thank the study team on the original *Accessibility in Science Labs* and *Making Practical Spaces Accessible* background papers and resource guides (see accessiblecampus.ca) for their collaboration on those projects; in particular, we thank Tina Doyle for her helpful insight and discussions in the early stages of this work. We also acknowledge those who have contributed to the research encapsulated herein throughout the course of this effort.

We thank the Special Initiatives Unit of the Council of Ontario Universities for their gracious permission to expand their original projects into this work, and the National Educational Association of Disabled Students for their initial leadership of the discussion around disability and the sciences.

We acknowledge the support of the Canadian National Institute for the Blind in the production of accessible formats of this book, and the advice provided by the World Blind Union on the importance of this work to the ongoing dialog around science education for persons with disabilities as a human rights issue on the international stage.

Finally, we thank our editors at Elsevier Publishing, Mary Preap and Fenton Coulthurst, for their diligence and collaborativeness in seeing this project through to completion.

Mahadeo A. Sukhai
Chelsea E. Mohler
Toronto, ON, Canada
May 2016

I am extremely grateful to my family and friends for their unwavering belief in my abilities and reaffirming words of encouragement. To my parents, I am extremely grateful for your interest in, and support throughout, this journey.

- C.E.M.

The ability to tell this story would not have come about without all those who contributed to my professional success in the sciences over the years. That long list of individuals includes my current and former supervisors; the students and postdoctoral scholars I have trained and worked with throughout my career; the best and brightest of my educators; my labmates, and in particular, those who provided technical assistance throughout my sojourn in the bench sciences; and the friends and family who encouraged and supported me throughout my development as a scientist and researcher.

- M.A.S.

Introduction

A framework for dialog around disability and the sciences

Access to education for youth with disabilities is enshrined within the UN Convention on the Rights of Persons with Disabilities as a fundamental human right (Article 24 of the Convention). Education in the sciences often poses a significant challenge to students with disabilities for a variety of reasons, both intrinsic and extrinsic to the student, the educator, and the education system as a whole. Lack of access to science education is therefore a human rights concern, as this significantly and negatively impacts the potential for full participation in society for youth with disabilities as well as restricts on a fundamental level their possible career paths—in essence, taking away or limiting the right of the youth to choose their own career trajectory.

With this book, our aim was to create a much-needed resource that provides required background and frameworks around barriers faced by students with disabilities in science, as well as suggest possible solutions for the removal of those barriers. This is not intended to be a "how-to" guide, or a manual of pedagogy or accommodation. Instead, this book is a guide to understanding the right questions to ask, the right conversations to have, and the right people to have them with.

We have four intended target audiences for this book: Educators at the postsecondary level (including faculty, teaching assistants, course instructors, and supervisors working with students with disabilities in the sciences, as well as policymakers in higher education); educators in the kindergarten through secondary school education system (including teaching and special education staff); students with disabilities in, or hoping to pursue, science, technology, engineering, or mathematics disciplines, at all levels of education; and, accommodation specialists in educational institutions, nongovernmental organizations, community service providers, and governments who work with students with disabilities in the sciences.

We have brought an international—even global perspective—to the preparation of this volume. Although the material is written as though it applies directly to the postsecondary education setting in the developed world, we believe that the significant human dimensions required to create a true culture of accessibility in the sciences for persons with disabilities are applicable globally, across all nations of the developed and developing world. Where appropriate and applicable, we stress this perspective throughout the book.

We hope the members of each target audience take something significant away from this work. For example:

- We anticipate that *postsecondary educators* will better understand the nature and crucial importance of their role in working with students in the sciences and with service providers to overcome systemic barriers faced by students with disabilities in the sciences.
- We hope that *educators in the K-12 system* understand the applications of the principles of open communication, principles of inclusive teaching, mentorship, and accommodation practices from kindergarten to grade 12, to foster inclusion and participation of students with disabilities in science courses.
- We aim for *students with disabilities* to appreciate the potential for, and requisites of, their successful participation in science courses and programs, as well as in science-based careers globally.
- Finally, we aim to provide the necessary background for *accommodation specialists working with students with disabilities in the sciences*, so that they may lead relevant and effective discussions in support of students with disabilities in the sciences.

The book is divided into seven sections. Section 1, Students with disabilities in the sciences, provides an overall introduction to the international landscape of science education for students with disabilities: Chapter 1, The landscape for students with disabilities in the sciences, focuses on the systemic concerns that impact the success of diversity, accessibility, and inclusion initiatives in the sciences, and provides a framework for key discussion points throughout the book. Chapter 2, Accessibility and science, technology, engineering, and mathematics—the global perspective, sets the global human rights framework, and highlights issues around cultural perceptions of disability and the international mobility of scientists.

Section 2, Barriers faced by students with disabilities in the sciences, of the book discusses barriers to education and training in the sciences for students with disabilities: Chapter 3, Barriers faced by students with disabilities in science laboratory and practical space settings, focuses on barriers external to the student (e.g., resources, funding, accessible formats, technology, and attitudinal barriers), while Chapter 4, Student perspectives on disability—impact on education, career path, and accommodation, focuses on intrinsic barriers, as defined by the student's lived experience growing up. Chapter 5, Key role of education providers in communication with students and service providers, discusses disclosure and accommodation from the student point of view.

Section 3, Student as educator, delves into the student perspective in more detail: Chapter 6, Disclosure in the sciences, reflects on the rationale for a model for action and positive change in science education on the part of the student: advocacy, communication, and trailblazing. Chapter 7, Student as ACTor—recognizing the importance of advocacy, communication, and trailblazing to student success in STEM, highlights the impact of science education and training on mental health and well-being for students with disabilities, while Chapter 8, Mental health and well-being for students with disabilities in the sciences, looks at the value of peer-support networks and where to find them. Finally, Chapter 9, Peer-support networks, discusses the importance of networking to career development of students with disabilities in the sciences.

Section 4, Student as learner, focuses on the traditional learning environment—the science lab and the classroom—and the importance of accessibility in learning in the sciences. Chapter 10, Essential requirements and academic accommodations in the sciences, discusses the concept of "essential requirements" as they relate to science education. Chapter 11, Universal design for learning, and Chapter 12, Inclusive teaching practices, offer a discussion of universal design for learning and inclusive teaching practices, respectively.

Section 5, Students as mentees, trainees, and leaders, focuses on mentorship and supervision. Chapter 13, Faculty mentorship of students with disabilities in the sciences, discusses the crucial role the educator plays in ensuring student engagement and success in learning in the sciences. Chapter 14, Faculty supervision of students with disabilities in the sciences, delves into the importance of faculty mentorship for students with disabilities, while Chapter 15, The student in a leadership, mentorship, and supervision role, examines the role of the student–supervisor relationship in science training for students with disabilities. Finally, Chapter 16, Leveraging professional development and networking opportunities, provides an inflection on the discussion by looking at leadership, mentorship, supervision, and collaboration roles students with disabilities may take on during the course of their training in the sciences.

Section 6, Accommodating students with disabilities in the sciences, provides an in-depth discussion around accommodation of students with disabilities in the sciences. Chapter 17, Accommodating students with disabilities in science laboratories and in fieldwork; provides an overview of the topic and presents a framework for faculty and accommodation specialists to work within. Chapter 18, Human accommodation—laboratory/technical assistants in the sciences, examines human technical assistance, while Chapter 19, Mainstream technology as accessibility solutions in the science lab, looks at adapting mainstream technology in the science lab setting. Chapter 20, Assistive technology, discusses assistive technology, Chapter 21, Accessible formats in science and technology disciplines; reviews accessible formats, and Chapter 22, Simulation learning, describes the thought process around simulation learning. Finally, Chapter 23, Physical access in science laboratories, discusses physical accessibility and safety in the science lab.

In the concluding section, Chapter 24, Practicum placements, highlights practicum placements and Chapter 25, General principles of designing accessible learning environments in the sciences, offers some general considerations around accessibility and universal design in the sciences.

Each chapter is framed with a first-person narrative from one of the authors (Dr. Sukhai), unless otherwise identified, followed by a set of key learning objectives and definitions. The body of the chapter provides an in-depth discussion of the chapter topic in a nuanced, yet nontechnical manner. The book is intended to be accessible to all target audiences highlighted above, and strives to present a cross-disability, cross-disciplinary, perspective on accessibility at all levels of science education and training. While we have designed the chapters to be read in a logical manner, and there is a specific flow to the information content, each chapter may also be read independently and in any order.

The following key themes are found within the text:

1. Students, faculty, and disability services staff will all find critical learnings throughout each chapter of this book; while the source material is the same for all audiences, the specific learned outcomes are unique to each group in different situations.
2. We argue for creativity, cooperation, and collaboration among all target audiences in order to create a truly accessible, universally designed learning environment for all phases of science, technology, engineering, and mathematics training.
3. Conditions and circumstances naturally evolve, based on a person's disability, accommodation needs, program, field of study, and research; it is therefore critical to meet such changing circumstances with flexibility and responsiveness.
4. Students, faculty, and disability services staff all need to be open-minded, each in their own way, with respect to accessibility and disability in the sciences.
5. We argue for the use of this book as a resource to demonstrate to students, staff, and faculty that they are not alone in attempting to evolve solutions to accessibility issues in the sciences; many precedents for discussion and problem-solving exist in the field.
6. There are many myths and misconceptions surrounding the accessibility of STEM education and the inclusion of persons with disabilities in the sciences, which do not hold under rigorous and critical evaluation.
7. Finally, and most globally, we argue for the existence of champions in the field—successful students, engaged faculty—who will advocate for greater inclusion of students with disabilities in STEM, and for a more universally designed science learning environment.

It is our collective hope—that of the authors, and all the collaborators who we have engaged on this project—that this book will provide a valuable entry point for students, faculty, educators, and accommodation specialists in addressing the fundamental accessibility limitations inherent today in science education, in order to create a culture of true accessibility in the sciences in the years to come (Ontario Ministry of Education, 2013).

Dr. Mahadeo A. Sukhai, PhD
Chelsea E. Mohler, MSc (OS)
Toronto, ON, Canada
May 2016

Part I

Students with Disabilities in the Sciences

The landscape for students with disabilities in the sciences

Chapter Outline

Introduction   4
Some key terms   5
Exclusive, segregated, integrated, and inclusive education   6
Differences between undergraduate and graduate laboratory environments   7
Work in the Academic Environment   8
Participation of students with disabilities in the sciences   8
Glass ceilings in the STEM training pipeline   10
What is a "culture of accessibility?"   11
Definition of practical space environments   12
Application of best practices across disciplines   12
Previous forays in accessibility and STEM   13
Conclusion   14

I grew up wanting to be a scientist.

This was a dream that was ignited within me when I was four, a partially sighted child in the Caribbean. In that time, in that place, it must've seemed an impossible dream to those I cheerfully announced it to.

That didn't matter to me– after all, what four year old (disability or no) cares for or about impossibilities?

On my way to the fulfillment of that dream, I navigated difficult transitions in emigrating to North America, in transitioning into high school, into university, between scientific disciplines as an undergraduate student, into my Master's, into my Doctoral program, and then into what has become a very varied career. On my way to the fulfillment of that dream, I broke records for the youngest high school student in my area, and became the youngest student to attend my university.

I didn't set out to do those things in the beginning.

I set out to be a scientist.

The best one I could be, in fact – and the rest of it was incidental.

When I was young, the concept of a "glass ceiling" meant nothing to me. If you asked me, I would've said you found them in greenhouses and they were good for keeping in the warmth.

As I began to work more in the spaces surrounding higher education and diversity, I learned and then understood the term

Creating a Culture of Accessibility in the Sciences. DOI: http://dx.doi.org/10.1016/B978-0-12-804037-9.00001-2

more intimately. And I realized I had, in fact, in my education and my career, plowed my way through quite a few of them. I also realized that a lot of other people hadn't ever made it that far – or, perhaps, hadn't tried.

I wondered about that – after all, I didn't think I was all that special. I'd taken advantage of opportunities that had come my way, and I have had phenomenal teachers over the years, but surely others had too?

The barriers I have faced in my training and career in the sciences are, I have learned, typical for students with disabilities in these fields – in some ways, that's a bit sad, because I'd begun my training in the sciences well before the current cohorts of students. We will talk about those barriers – about educator attitudes and available resources, about the gatekeeper function and the assumptions people make, about the value and importance of good educators and good mentorship, and about our own attitudes and perceptions as people with disabilities learning the sciences.

These are all systemic issues to some degree, to be sure – but they are also eminently solvable, and, for me, they were solvable because of people who were willing to work with me to talk them through and to figure out where and when and how to push.

That was how the glass ceilings broke for me.

I think, for those of us in fields where we are not expected to be, or to succeed, the most powerful message we can receive is one that says "We are not alone." And, in reality, we are not – there is an increasing number of people with disabilities moving through training in the sciences, and this number will only increase over time.

I grew up wanting to be a scientist, never dreaming where that career would eventually take me.

I know now that others have that dream also.

Everyone deserves to live it, as I have.

Introduction

Why Write a Book Like This?

Education and employment are enshrined as rights of persons with disabilities in the United Nations Convention on the Rights of Persons with Disabilities (CRPD; Articles 24 and 27). The right to full inclusion and participation in society includes the opportunity to obtain an education in all disciplines important to the world's progress today, including the sciences.

Many of the science-based careers in today's evolving economy require at least the completion of a first-year university chemistry course (McDaniel, Wolf, Mahaffy, & Teggins, 1994). Furthermore, science is the basis of an increasing proportion of workforce opportunities (Hilliard, Dunston, McGlothin, & Duerstock, 2011). In addition to the science requirements for those pursuing careers in science, technology, engineering, and mathematics (STEM), it is typical of postsecondary institutions to require nonscience majors to complete at least one science credit, aimed at educating the nonscientist (Pence, Workman, & Riecke, 2003).

As opportunities in technical and medical fields continue to grow, all students— including persons with disabilities—need a strong education in science to achieve their career goals. Additionally, learning to think with scientific reasoning teaches critical thinking and allows for even nonscientists to learn new ways of teaching themselves. A strong philosophical argument in favor of inclusion also suggests that learning critical and analytical thought—one of the foundational essential requirements of any scientific discipline—should not be restricted on the basis of disability.

"Active experiences," where students engage in all aspects of laboratory activities, are critical to success in science-based careers, and students with disabilities are no exception. The underrepresentation of students with disabilities in STEM can be attributed to limited exposure to the sciences and a lack of teacher training in inclusive teaching practices (Moon, Todd, Morton, & Ivey, 2012).

Exposure to science education contributes to the development of an interest in the sciences. Young people with disabilities who experience special education or segregated schooling environments often lack exposure to science and may never have the opportunity to develop this interest. A lack of teacher training around appropriate accommodations in the sciences, particularly in the laboratory environment, will create barriers for students with disabilities (Moon et al., 2012). This extends to educators at the postsecondary level, who are in a position to foster the development of students with disabilities in lab-based science courses, curricula, and programs of study.

One answer to the question we pose, then, presents itself as the need to fill a knowledge gap. Lack of educator and service-provider awareness is a significant barrier to the full participation of students with disabilities in the sciences. Indeed, the student's development and growth will also suffer due to his or her lack of knowledge of role models and precedents. This book is intended to be a significant first step in closing that gap.

Some key terms

Before proceeding further in this book, it is important to make sure that we, the authors, and you, the reader, are using the same language to describe disability and education in the sciences.

STEM fields are considered to be any field in the sciences (including physical sciences, life sciences, and basic medical sciences), along with technology, engineering, and mathematics disciplines. Interdisciplinary and multidisciplinary fields are included in our discussion here.

Disability is "...an umbrella term, covering impairments, activity limitations, and participation restrictions. An impairment is a problem in body function or structure;

an activity limitation is a difficulty encountered by an individual in executing a task or action; while a participation restriction is a problem experienced by an individual in involvement in life situations" (*World Health Organization*, 2016). Disability is considered "…a complex phenomenon, reflecting the interaction between features of a person's body and features of the society in which he or she lives. Overcoming the difficulties faced by people with disabilities requires interventions to remove environmental and social barriers" (*World Health Organization*, 2016).

"In human rights terms, *accommodation* is the word used to describe the duties of an [educator], employer, [or] service provider … to give equal access to people who are protected by [Human Rights legislation]," including persons with disabilities (*Human Rights Legal Support Centre, Government of Ontario, Canada*, 2016). Meanwhile, *Accommodation needs* are the "tasks and … functions that a person with a disability cannot fully perform without some type of accommodation" in the context of their course, program, or discipline (*Work Without Limits*, 2016).

Reasonable accommodation is any change to a job, the work environment, or the way things are usually done that allows an individual with a disability to apply for a job, perform job functions, or enjoy equal access to benefits available to other individuals in the workplace (*US Department of Labor, Office of Disability Employment Policy*, 2016). A similar concept exists within all levels of education. Educators and employers are required by law within their jurisdictions to provide reasonable accommodation to qualified students and employees with disabilities, unless doing so would impose an undue hardship. Finally, *undue hardship* is "an action requiring significant difficulty or expense when considered in light of … the nature and cost of the accommodation in relation to the size, resources, nature, and structure of the [employer's or school's] operation. Undue hardship is determined on a case-by-case basis" (*Frequently Asked Questions, Americans with Disabilities Act 1991*).

Exclusive, segregated, integrated, and inclusive education

In the developed world, we are accustomed to thinking about education of students with disabilities as *segregated*, where students with similar disabilities are grouped together and taught separately from their nondisabled peers (e.g., residential schools for blind or partially sighted children); or *integrated*, where students with disabilities can attend a mainstream school, but are attached to special education classes and/or provided with itinerant teachers to support their learning. In the developing world, education systems may *exclude* students with disabilities entirely, and may even argue that this is for the student's own good, because the school may not be set up to provide the appropriate education for him or her. That argument may also apply to specific courses or programs within the education context, particularly the sciences.

Each of these three types of education can significantly and negatively impact a student's engagement with the sciences, as well as access to science courses. Ultimately, students may be directed or encouraged away from science courses entirely, or their participation limited, due to educator awareness, attitudes, and knowledge, or to available resources for accommodation. In some cases, the choice may be seen as between participating in science courses, which the student may be perceived as deriving no benefit from, and taking life skills courses or Braille courses. However, this is a false choice. A basic instruction in the sciences, and basic scientific literacy, are required for participation in society today. Lack of access to science education is fundamentally a human rights issue, as this restricts students' rights to choose their educational interests and career path.

Throughout this book, therefore, we argue for *inclusive* education in the sciences, whereby the environment of learning and the culture within the discipline itself is made to be welcoming to students with disabilities, and where the learning environment evolves to meet the needs of the learners. By extension, we also argue for an inclusive work environment in the sciences.

Differences between undergraduate and graduate laboratory environments

The concepts we discuss in this book with respect to the educational environment in the sciences are relevant and applicable to both the secondary school setting and to postsecondary education. In postsecondary education, graduate education is much different than undergraduate education. Students are expected to be much more independent in their laboratory research, and they are expected to take on leadership roles in planning and assisting with courses. They are now becoming an expert in their own area, and should be ready to be questioned and challenged. In the lab setting, students are often required to work as a part of a research team, and could potentially be a part of a larger research project (having their own portion) or have their own project they're responsible for carrying out. Inquiry-based study involves students in a "discovery process" (Weaver et al., 2008), which in science can include designing a research problem, conducting appropriate experiments, analyzing data, and presenting the results in a meaningful context. This is in contrast to "cookie cutter" teaching laboratories, wherein students follow a set of instructions, usually designed for them by the educator, generating predictable results that confirm the proposed hypothesis, demonstrating a scientific concept, or reinforcing course content in some manner (Luckie et al., 2012). The large majority of undergraduate teaching laboratories follow the "cookie cutter" format, but it is worthwhile to ask, is this the best way to make use of that laboratory time?

Undergraduate laboratories are a perfect opportunity for students to practice being a scientist; and scientific research is hardly ever black and white like undergraduate teaching labs are sometimes designed to be (Hartman, 1990). Research requires consideration of the literature, careful planning, critical thinking, and communicating the analysis in a meaningful way—skills we should be teaching our students in any science course. Research also often involves failed protocols, results that are difficult to interpret, and frequent trouble-shooting—scenarios we should make students aware of in order to prepare them better for graduate school and the working world.

While an undergraduate honors thesis will have laid a foundation for independent research, there will be a variety of other, potentially new, responsibilities in a graduate research setting. This may overwhelm new grad students, who may benefit if they take advantage of campus resources and departmental mentors.

Work in the Academic Environment

Undergraduates often enter graduate school never having taught a lab, given a lecture, or worked on a research project. It is important for students to experience these opportunities, as these add critical professional development skills for the postgraduation job market (Hartman, 1990).

As graduate students, they will be in a unique situation of being both teachers and students. Teaching assistantships (TAs) are a great way to develop skills and make extra money; however, new grad students may be hesitant to pursue a position or may be worried about their abilities to manage a class on their own. There are valuable resources through university and college campuses' centers for teaching and learning that can help them prepare for the classroom.

Advising undergrads who are interested in grad school to keep an eye out for research assistantships is a good idea, as this is a great way for them to build their own researching skills and confidence. Students may find opportunities in research and teaching by speaking with their teaching faculty, supervisors, academic advisors and/or thesis committee members. Getting involved with additional projects can create opportunities to build strong networks across disciplines and departments, and help develop research management skills for their own work.

Participation of students with disabilities in the sciences

We know that, today, the relative proportion of students with disabilities in STEM-related fields is low. Data from the United Kingdom (the Campaign for Science and Engineering's *Improving Diversity in STEM* report, 2014), the United States (the National Science Foundation's *Women, Minorities and Persons with Disabilities in Science and Engineering: 2013* report; Moon et al., 2012; Pence et al., 2003), and Canada (Chambers, Sukhai, & Bolton, 2011, among others) tell a striking, if poignant tale of underrepresentation:

- In the United Kingdom in 2012, there were approximately 4000 students with disabilities enrolled in all STEM disciplines in postsecondary education, at all levels;
- Only 5% of graduates with disabilities in the UK identified as being in STEM programs, compared to almost 20% of the nondisabled student population;
- Persons with disabilities were significantly less likely to work in STEM-related fields than their nondisabled peers;

- A total of 350 doctoral degrees in science and engineering were earned by persons with disabilities in the United States in 2009 (against an overall population of 350 million);
- Students with disabilities in Canada were least likely to pursue basic sciences studies (programs in life and physical sciences) in postsecondary education, with a significant drop-off between undergraduate (baccalaureate) and graduate (master's and doctoral) education.

(Despite our best efforts, we were not able to obtain data from the European Union, Australia, or South Africa, or other major jurisdictions, suggesting that these data are not routinely collected—which, itself, lends credence to the notion that students with disabilities in the sciences are altogether infrequent.)

We know some of the reasons for this lack of participation. Perceived and actual barriers play a large role in deterring students with disabilities from pursuing careers in the sciences (Hilliard et al., 2011). Students are often discouraged by educators and societal attitudes, from an early age, and within higher education, from pursuing laboratory-based science programs (Hilliard et al., 2011; Miner, Nieman, Swanson, & Woods, 2001; Moon et al., 2012). Access to financial aid and relative cost of accommodations pose another significant systemic barrier, although this is not unique to students with disabilities in STEM fields (Chambers et al., 2011; the Campaign for Science and Engineering's *Improving Diversity in STEM* report, 2014). We know too that science labs are relatively inaccessible environments, from a physical and navigation perspective.

If all this is true, why invest resources into creating a book that aims to introduce a culture of accessibility into the conversation around STEM education?

Another answer, then, is three-fold: First, participation of students with disabilities in STEM-related fields is low, but it is not zero—a number of students with disabilities have succeeded in attaining doctorates in multiple science and engineering fields. This generation of scientists with disabilities forms a group of role models for current students; the American Association for the Advancement of Science maintains one repository of success stories online at www.sciencecareers.org, while Canada's National Educational Association of Disabled Students maintains another repository at www.neads.ca. Additionally, students with disabilities pursing advanced STEM training are increasingly reaching out to one another through academic conferences, organizations of persons with disabilities, social media, and formative professional networks, such as PhDisabled (https://phdisabled.wordpress.com/), based in the United States, and Chronically Academic (https://chronicallyacademic.org/), based in the European Union.

Second, the participation rates of students with disabilities in STEM is on the rise. Data from the United Kingdom suggests that, with better design specifications and building codes, the number of students with physical disabilities entering the sciences has increased by 50% in the 3 years from 2009 to 2012. Meanwhile, participation of students on the autism spectrum over the decade from 2001–2002 to 2011–2012 went from 0 to more than 1000 students. Parallel growth was seen in the United Kingdom for students with mental health disabilities (the Campaign for Science and Engineering's *Improving Diversity in STEM* report, 2014). Furthermore, evidence suggests that as more students with autism spectrum disabilities are entering postsecondary education, they are being disproportionately drawn to STEM disciplines, particularly computer science and the basic sciences (Wei et al., 2013). Although reasons for this attraction have not been reported, a reasonable speculation might be that the natural rigor of the scientific method may be attractive to these students.

Third, technologies are changing, along with legislation, increased awareness, and evolving attitudes toward societal participation of persons with disabilities. While we still have a very long way to go, the concept of full inclusion and participation of persons with disabilities in society has gained significant traction—indeed, it is enshrined in the UN CRPD. Such changes translate into opportunities for people with disabilities, opportunities which historically have not existed before.

So, we begin to experience a world where students with disabilities may not only express their dreams of being a scientist, but where the tools and attitudes are there to make such dreams a tangible reality. It is for that reason that this book, and the concept of a culture of accessibility that it describes, are important and timely.

Glass ceilings in the STEM training pipeline

Conventional wisdom suggests that the scientific training pipeline takes in students coming out of secondary school and moves them through college or undergraduate baccalaureate programs, then on through master's and doctoral programs, into post-doctoral training and into jobs within the scientific workforce. In reality, this pipeline concept was coined in the context of the training ground for the professoriate, and in today's funding climate, is much more akin to a funnel—one where trained persons are "siphoned off" as they choose to pursue other career tracks.

(If one makes the conceptual leap that suggests that the desired outcome is any job that one is qualified to do in the scientific workforce, then the "pipeline" becomes more of a "pipe track" around a very large lake, with multiple spigots into that large body of water—although that might take the analogy too far.)

As with other underrepresented groups in STEM—women and visible minorities, for example—we can imagine a conceptual glass ceiling in the STEM training pipeline (Allen-Ramdial & Campbell, 2014). Indeed, we can imagine four such barriers:

1. The barriers that present themselves to students with disabilities at the point of entry into postsecondary education;
2. Those barriers that arise during the transition from one stage of postsecondary training to the next (including the transition into the postdoctorate);
3. Barriers that arise in moving into the labor market and workforce (which can be similar to transitional barriers); and,
4. Barriers to effective participation in STEM training that arise while the student is within the learning environment.

Many diversity and inclusion initiatives are focused on what are referred to as "supply side" approaches—i.e., approaches to increase the flow of students from underrepresented groups into the training pipeline. At best, these initiatives address the first of the four glass ceilings discussed above, without appropriately addressing the others. Indeed, "supply side" approaches are often limited in that they do not fully address the issues faced by students in their secondary school careers that may prevent them from choosing to pursue STEM training in postsecondary (see chapter 3: Barriers faced by students with disabilities in science laboratory and practical space settings). Akin to ways recently addressed by others for women and visible minorities (Allen-Ramdial & Campbell, 2014), we offer through this book a variety of approaches aimed at deconstructing the "glass ceilings" in the STEM training pipeline for students with disabilities by:

1. *Improving the preparation of students with disabilities moving through secondary school and into postsecondary education.* Many of the educational and accommodation barriers that exist in postsecondary also exist at the elementary and secondary school levels. The material in this book is intended to provide an appropriate background to science teachers, parents, and special education staff in working with and identifying accessibility solutions for their children and students with disabilities.
2. *Improving the preparation of students with disabilities moving through postsecondary education.* By describing the breadth and variety of learning environments, as well as the different faculty interactions and accommodation solutions, possible in STEM education at the postsecondary level, it is hoped that students themselves will be better prepared to navigate their learning environments and educational transitions.
3. *Improving faculty, service provider, and employer awareness.* The material presented herein, while intended to be of benefit to students, is primarily targeted at faculty, service providers, student-life professionals, career educators, and employers who currently, or are likely, to work with students with disabilities. Much of the background and strategies presented are intended to provide necessary knowledge as to the issues, barriers, and potential solutions to work through with students with disabilities in STEM.
4. *Improving accessibility of STEM education.* STEM education is a complex enterprise, with many discrete components. By working through a variety of learning environments and accommodation solution approaches with the reader, we intend that faculty, service providers, student-life professionals, and students will be prepared to engage in collaborative discussions around the student's participation in STEM education in an appropriate and accessible manner.

What is a "culture of accessibility?"

The STEM literature has focused for a long time on the physical accessibility of a science laboratory. As we will return to in Chapter 23, Physical Access in Science Laboratories, the physical structure of most laboratories is unwelcoming to persons with physical disabilities; at worst, it is inaccessible. Many laboratories are laid out as hard to navigate and visually obstructive environments. Lab spaces are often encumbered by high workbenches, inaccessible cabinets, and overcrowded fragile equipment (Hilliard et al., 2011). Furthermore, the STEM literature has not—until very recently (see Doyle, 2014)—developed a standard design for an accessible science lab in any field (Moon et al., 2012). Additionally, the literature provides little in the way of examples for accessible laboratory equipment used in the instruction of STEM courses at the postsecondary level (Moon et al., 2012).

The STEM literature has not substantively discussed other aspects of a campus-wide culture of accessibility in the context of the sciences, a concept we first introduced in our initial work in this area (Sukhai et al., 2014a; 2014b). Components of a culture of accessibility in STEM include faculty–student interaction, inclusive teaching practices, content delivery, and mentorship, among others.

However, the components of a culture of accessibility significantly impact, and may be relatively more important than, physical accessibility in defining a student's ability to participate fully in science labs at the graduate and undergraduate levels. These include teaching practices, the relationship between the student and faculty members, and the accessibility of content delivery methods and lab equipment. Furthermore, a focus on physical accessibility of the lab addresses barriers faced by students with physical and sensory disabilities, while potentially minimizing the challenges faced by students with "invisible," mental health, learning, or cognitive disabilities.

Definition of practical space environments

A significant focus of this book is the accessibility of science-related learning environments, and in practical spaces associated with, and oriented toward, STEM training. We define a "practical space" as a learning environment where students have the opportunity to engage in active learning and to demonstrate, through hands-on activities, the practical components of a given discipline. Some examples of practical spaces include basic science laboratories (c.f., Sukhai et al., 2014), occupational therapy and physical therapy laboratories, design studios, observatories, and archives and museums. Practical spaces are used to supplement classroom learning, and provide additional opportunities for students to work through practical scenarios typical of their respective disciplines.

The definition of practical spaces was crafted based on the authors' own experiences, and on the results of key informant interviews conducted for this project. Our review of the limited literature elicited, perhaps not surprisingly, a limited theoretical and practical framework from which to draw meaningful and useful conclusions. This highlighted the critical need to develop a comprehensive experiential synthesis for faculty and service providers working with students with disabilities in practical spaces.

Application of best practices across disciplines

An identified need for further research goes beyond describing barriers to science: it provides educators with resources and best practices that address all aspects of the accessibility of a science laboratory.

Most literature on the accommodation of students with disabilities (specifically visual impairments) in the science-lab setting has focused on chemistry education and/or accommodations at the secondary level. It is worthwhile to note, however, that the concepts of accessibility and universal design are translatable across multiple contexts and disciplines. Anecdotal evidence from students and young scientists with disabilities in various disciplines suggests that the accessibility of the science laboratory environment has a number of common themes. These include the importance of:

- Creativity in addressing academic accommodations, particularly with technology adaptations;
- A strong relationship, or "partnership," with faculty—either the course instructor/coordinator or the thesis supervisor, which includes both trust and support from both sides of the relationship;
- A flexible teaching approach; and
- Creativity and flexibility in meeting the essential requirements for a course, program, and discipline.

Previous forays in accessibility and STEM

One initiative designed to increase opportunities for students with disabilities is the construction of the Accessible Biomedical Immersion Laboratory at Purdue University in the United States. The purpose of this project is to enable persons with physical disabilities to access standard laboratory and safety equipment. Modifications to a science laboratory at Purdue University's Discovery Learning Research Centre were made to facilitate scientific research for students with disabilities in a learning lab environment (Hilliard et al., 2011). A similar initiative, the iScience (Integrated Science) laboratory, was undertaken at McMaster University (Canada) in the General Sciences program. This accessible lab opened in September 2013. The iScience laboratory is designed to teach numerous disciplines: life science, physics, chemistry, and so on. All utilities are in place in an integrated setting, with safety features fully considered as well.

Notwithstanding the barriers to accessing science laboratories for students with disabilities, there have been several initiatives undertaken to facilitate a positive learning experience for these students. Initiatives include the dissemination of publications and resources on the construction of accessible science laboratories and fostering mentorship opportunities. One such program that works to facilitate opportunities for students with disabilities is the Disabilities, Opportunities, Internetworking, and Technology (DO-IT) program at the University of Washington. DO-IT strives to increase the number of persons with disabilities pursuing careers in science. The program provides mentorship through a summer learning opportunity, and is host to an online community of scholars. Additionally, the DO-IT initiative is a hub of online resources for teachers, students, and faculty.

Another major initiative was led by the authors of this work (Sukhai et al., 2014a; 2014b), and eventuated the creation of the initial work that this book is based on. There, we created a pair of comprehensive background reports as well as a series of educational resource guides that would explain how to make all types of science laboratories accessible for students with disabilities at the undergraduate and graduate levels, and introduced to the field the concepts of "culture of accessibility" and "practical space" learning environments in the context of student learning in the sciences.

Here, we expand significantly on our initial work, by detailing the thought process around the principles of accommodation, as well as by providing additional insights into the application of various types of accommodation solutions to the scientific setting; by discussing the type and quality of the various learning roles a student with a disability engages in during his or her time in STEM training, as well as the interactions with faculty members in each of those roles; and by setting the conversation in the global context of the scientific research enterprise and STEM training.

Conclusion

The DO-IT (2011) program, based at the University of Washington, emphasizes that "academic preparation from an early age, self-advocacy, universal design of learning and work environments, and acceptance by educators, employers, and peers, [are] a recipe for success in STEM for individuals with disabilities." The intent of this book, and the material and concepts contained herein on making science laboratories accessible for persons with disabilities, is therefore to explore efficacious academic supports to address the decreased participation of these students in lab-based STEM fields (Street et al., 2012). The content of the book will support faculty in understanding that they play a key role in creating an inclusive laboratory experience. Furthermore, these resources are intended to enhance faculty knowledge of inclusive teaching practices, adaptive equipment, accessibility features, and accommodations in science laboratories, which would enable the successful participation of students with disabilities in the sciences.

Accessibility in science, technology, engineering, and mathematics—the global perspective

2

Chapter Outline

Introduction   16
Accessibility of science, technology, engineering, and mathematics (STEM)
   education: a human-rights framework   17
United National convention on the rights of persons with disabilites   18
Higher education in the context of the global landscape of disability rights
   legislation   19
Cultural perspectives about disability   20
Science as an international endeavor   21
Conclusion   22

Amanda's Story: By the time that I made my pilgrimage to Ireland I had known for more than a decade that I wanted to study abroad. At the time, I could not have put my finger on exactly what the appeal was for me. I was drawn to the imagery of very old growth forests and the ancient ruins which occupied the landscape from coast to coast. Specifically, I had a deep sense of ages gone by, which I felt in the physical structures of the ruins which pepper many lush green Irish fields. Like most Americans, my ancestors were immigrants and Irish was just one of the nationalities which contributed to my lineage. I was curious to see the land which some of my direct ancestors had inhabited. My appreciation of ancient things was extended to DNA that same year. While carrying out genetics based studies as a graduate student at Trinity College Dublin, I began reading about the resident Ancient DNA laboratory and its research on medieval age animal products such as leather and parchment. While I was not strongly interested in any of the domesticated species which this laboratory studied, my interests were very much piqued by the prospect of population genetics and evolutionary history.

While in Europe, I took several opportunities to travel and see iconic structures and landscapes. I visited Paris, Portugal and the UK during this time. While experiencing the cultures was thrilling, I most wanted to feel the age of these places. My fondest memories are of sitting on the rocks at Giant's Causeway and starring out into the ocean, crouching beside the megalithic portal known as Brownshill

Creating a Culture of Accessibility in the Sciences. DOI: http://dx.doi.org/10.1016/B978-0-12-804037-9.00002-4

dolem and walking through the Neolithic monument at New Grange. More than just ancient peoples' genes persisted into the modern day, and these monuments were testimony to these remarkable feats.

Mahadeo's Story: They tell us, as young scientists, to be mobile – to move around, to gain new experiences and an appreciation of different ways of doing things. I never did that – indeed, for a variety of reasons, I chose to be a bit more sedentary than Amanda. I stayed put, and let the world come to me.

Mine was the route less traveled, literally and figuratively, since most of my peers in my Master's and Doctoral programs eventually did go to other places and experience other ways of doing things. A mentor of mine once encouraged me to not think negatively about my choice – his argument was that, being embedded within one of the largest universities in the world, in one of the most multicultural cities, I was exposed to different ways of doing things and the full diversity inherent in the scientific enterprise, merely by walking across the hall.

He was right.

Both paths – mine and Amanda's – are equally valid for all young scientists, with or without disabilities, and I know of a few students and peers of mine who, for their own reasons, chose to stay put after completing their doctoral degrees.

Irrespective of the path we choose, as young scientists, eventually we will run up against different thought patterns that arise from different cultural backgrounds. Coming from the developing world, and having been exposed to the cultural attitudes (and availability of resources for persons with disabilities) in the Caribbean, the North American cultural norms around disability and self-advocacy were a bit of a shock. I internalized them, but if I were to be totally honest, they don't always fit comfortably.

One thing about the increasingly global nature of science is how many internationally trained peers, collaborators and faculty members I encounter – and how many cultural attitudes about disability I experience as a result. I'm maybe a bit more sensitive to those different attitudes, and the nuances of how they influence people's interactions with me, because of my own upbringing.

Introduction

It is currently estimated that approximately 15% of people worldwide are living with some form of disability (World Health Organization, 2015). This is a higher percentage than previous reports, which estimated a 10% disability rate, and is due to an aging population, the rise of chronic diseases and disabilities worldwide, and improvements in methodologies used to measure disability. While systemic barriers may currently prevent a large proportion of children with disabilities from attaining primary and/or secondary education in developing countries, as technology and infrastructure continue to advance, more and more students with disabilities are able to pursue dreams of a postsecondary education in fields of their choosing. The international legislative framework around disability, as well as cultural attitudes and perceptions of disability, significantly influence postsecondary institutions' infrastructure around disability accommodation as well as faculty, educator, and service-provider attitudes. Additionally, due to the high international mobility rate of students, research trainees, and faculty in the sciences, cultural competency and cultural attitudes around disability are important factors in student–faculty and peer-to-peer interactions. This chapter briefly highlights some of the key issues in considering disability in the context of the globalization of science.

Accessibility of science, technology, engineering, and mathematics (STEM) education: a human-rights framework

Basic numeracy and scientific literacy are required the world over for full participation in society. Irrespective of how far one chooses to advance in science, technology, engineering, and mathematics (STEM) education and training, access to science courses during primary, secondary, and tertiary education ultimately impacts our ability to participate fully in society. Restricting access to or participation in STEM courses, at any stage of one's education path, will therefore have a negative impact on the ability of youth with disabilities to participate fully within their communities and in society at large.

A further challenge also arises in the context of education path and career choice. Only those technical disciplines whose bonafide occupational requirements may necessitate a person to perform specific tasks, which are impacted by disability, without accommodation or assistance may be considered as "out of bounds" for a youth with a disability making his or her career choice (we revisit this topic in the context of the educational setting in chapter 10: Essential requirements and academic accommodations in the sciences). Restricting access to and participation in STEM courses and programs artificially restricts students' rights to choose their career path, limiting them to opportunities that do not require working knowledge of the sciences or mathematics. Even within a variety of non-STEM careers, some specialties require working knowledge of scientific and mathematical concepts—e.g., a lawyer specializing in patent and intellectual property law benefits from some scientific training (Azzopardi, Johnson, Phillips, Dickson, Hengstberger-Sims, Goldsmith, et al., 2010).

Thus, a lack of access to, or restriction of access to and participation in, STEM education at any level becomes a human rights concern. This lack of accessibility translates to a restriction in choice of career paths and jobs, a concomitant lack of employment and mobility, and, more fundamentally, a limitation imposed on a person's right and ability to participate fully within their community and within society as a whole.

In the context of this human-rights framework, educator and accommodation specialist training on how best to work with students with disabilities in STEM disciplines becomes critical. In the developing world, especially, there is a shortage of teachers with the appropriate training and qualifications to work with youth with some disability types (e.g., those who are blind or partially sighted), and fewer still have both the training to work with students with disabilities and the qualifications to teach STEM disciplines.

In all parts of the world, and at all levels of education, this critical shortage of educator and accommodation specialist training and background knowledge can coexist with a tension to ensure that youth with disabilities have appropriate life-skills training. As a result, in some education environments, an artificial choice is constructed: Life-skills courses, or Braille instruction, for example, are offered as necessary replacements for STEM education. We can envision, in the context of the human-rights framework just described, the impact of these tensions and choices, on youth with disabilities' participation in STEM, and in society at large.

United National convention on the rights of persons with disabilites

The United Nations Convention on the Rights of Persons with Disabilities (CRPD) was ratified by the UN General Assembly in 2008. More than three-quarters of the world's nations are signatories: At the time of writing, 159 counties have signed the Convention, with 156 having ratified it through their parliaments or national assemblies. 92 of 156 counties have also signed the Convention's Optional Protocol, with 86 of those having also ratified it (UN.org). As a result of the significant uptake of the CRPD, most countries worldwide either have signed or ratified the CRPD, or have antidiscrimination language in their national constitutions, or both (Disability Rights Education and Defense Fund: www.dredf.org).

Participation as a signatory to and ratification of the Convention as well as its Optional Protocol has been most prevalent in the countries of Latin and South America and the Caribbean (www.UN.org); in the countries of the European Union, where 28 nations have ratified (European Union Agency for Fundamental Rights: http://fra.europa.eu/); and West and South Africa. Many additional countries—including Canada, the United States, Russia, India, and China—have signed or ratified the CRPD, without the Optional Protocol. Of particular relevance to trainees with disabilities in the sciences, most countries with well-developed research infrastructure have signed or ratified the CRPD.

Within the language of the CRPD itself (www.un.org/disabilities/convention/conventionfull.shtml), Articles 18 ("Liberty of movement and nationality"), 24 ("Education"), and 27 ("Work and employment") are most relevant to the training and employment of students with disabilities in their fields of interest (in particular, the sciences, given the topic of this book). Specifically, these articles provide an international foundation for prohibiting discrimination and protecting the rights of persons with disabilities through all phases of their education (including postsecondary), and into their employment. Additionally, Article 18 of the Convention also prohibits discrimination in the context of cross-national and international mobility of persons with disabilities, something that is of particular relevance to all research trainees, particularly those in the sciences.

Of secondary relevance, but still important to the international human-rights framework around participation in STEM, are Articles 9 ("Accessibility"), 21 ("Freedom of expression and opinion, and access to information"), and 30 ("Participation in cultural life, recreation, leisure, and sport"). Article 9 reinforces the importance of physical and informational accessibility of facilities, including educational institutions, as well as the importance of provision of the appropriate accessibility resources, including human assistance, to support persons with disabilities attending school. Article 21 acknowledges the importance of a variety of accessible communications methods (including languages such as Braille and Sign Language) in fostering the participation of persons with disabilities in society, including in the educational environment. Article 30 highlights the importance of persons with disabilities having the opportunity to pursue their hobbies and interests, including those that may be STEM- or education-related.

In addition to the CRPD, approximately 45 countries worldwide have enacted specific disability legislation at the national level (www.dredf.org). Other countries, like Canada, have regional or provincial disability legislation (e.g., the *Accessibility for Ontarians with Disabilities Act*), but as yet no national legislation. National disability or antidiscrimination acts vary in their aims, definitions of disability, terminology, specific language, and compliance protocols, and navigating these complexities can be a challenging exercise. Of note, these pieces of legislation exist in tandem with both the CRPD and differing cultural perspectives on disability around the world. The CRPD is considered to be the international framework that countries' national legislation must be in agreement with; indeed, national disability legislation is viewed as a country's specific manifestation of the CRPD.

Higher education in the context of the global landscape of disability rights legislation

Many universities and colleges, or their equivalent institutions around the world, will deploy specific offices to provide services to students with disabilities. These offices may take different forms, and may operate under significantly different policies based on their native legislative and cultural frameworks (United Nations Educational, Scientific and Cultural Organization, 2005).

In more resource-poor areas, where a necessary and critical focus remains on countering cultural attitudes around disability and ensuring that students with disabilities successfully complete primary and secondary school, participation in postsecondary is neither expected nor emphasized. Individual students with disabilities seeking to pursue postsecondary studies in any field, let alone the sciences, find themselves trailblazers in educational settings unprepared to work with them.

In more resource-intensive countries, students with disabilities must contend with cultural attitudes around disability as well as the policies and practices in place at the institution, which are in turn fostered by the legislative frameworks around disability and higher education in that country. In our investigation, students with disabilities faced many of the same accommodation challenges in the sciences, irrespective of their training site (North America, Europe, or elsewhere): Understanding the nature of the accommodations required by students in order to succeed, and fostering their implementation, proved to be common barriers at an international level, and our discussions within this book are reflective of that observation.

It is also worth noting that there is a level of commonality to the discourse around disability, integration, and inclusion that is fostered by youth and young professionals with disabilities globally and in regional and national forums. Commonly, youth focus on the importance of education and employment in their dialogues through a range of youth-oriented and professional organizations such as Canada's National Educational Association of Disabled Students; the Canadian National Institute for the Blind's National Youth Council; the Foundation Fighting Blindness' National Young Leaders Program (also in Canada); the National Association of Blind Students in the United States; PhDisabled (also in the United States); Chronically Academic in the European Union; and the World Blind Union's Youth Engagement Committee, among many others.

Cultural perspectives about disability

Cultures around the world perceive and treat disabilities very differently. Many of these differences may arise from religious beliefs (e.g., the concepts of reincarnation and ancestral sin) or social mores (e.g., disability as shameful, the imperative toward social conformity, or cultural perspectives on the behavior of boys compared to girls; Al-Salehi, Al-Hifthy, & Ghaziuddin, 2009; Baker et al., 2010). Different disabilities are perceived differently, based on the nature of a child's needs and/or behaviors, with particular culture-specific differences arising for learning disabilities, mental health disabilities, and developmental disabilities (e.g., autism spectrum). Indeed, cultures may view disabilities in three ways: By their cause, by their effect on societally valued attributes, and by the status of the person with a disability as an adult (Groce, 1999).

The impact of a person's native culture and upbringing on their perspectives around their own disability(ies) is highlighted in chapter 4, Student perspectives on disability—impact on education, career path, and accommodation; here, we discuss the impact of cultural perspectives on disability on the interactions that a student with a disability may have with his or her peers, faculty, and potential employers.

In the context of interactions within the educational system, the cultural perception of disability through its effect on societally valued attributes is important, as is perception based on the person's adult status. A person with a learning disability, for example, may have more difficulty interacting with educators and peers in a culture where intellectual prowess is narrowly defined, while a person using a walker or other mobility aid might find more challenges in a culture where physical strength is prized. Cultural, community, and religious expectations around disability play out as youth with disabilities are often "streamed" into education paths thought to be most suitable to their disability (as opposed to their skill set, aptitudes, and interests). It is important to note that, in multicultural societies where native cultural perceptions and practices are reinforced in the microcosm of any given cultural community, these expectations may also be found, and can confound interactions among parents, students with disabilities, educators, and accommodation specialists at all levels of education.

The specific demonstration of different cultural perspectives on disability by either the parents or students will be different than if exhibited by educators and accommodation specialists, based on the nature of the student–educator relationship. For example, when the educator or faculty member has strong cultural attitudes around disability, the power dynamic between faculty and students could translate into strong negative attitudes around student participation, obstructionist behavior on the part of faculty, or demonstration of the "gatekeeper function" (see chapters 3: and 10: Barriers faced by students with disabilities in science laboratory and practical space settings and Essential requirements and academic accommodations in the sciences for a more full discussion of this term). This dynamic, which is compounded by cultural attitudes, can also play out in a faculty member's mentorship of a student with a disability (see chapter 14: Faculty supervision of students with disabilities in the sciences), or his or her supervision of a student with a disability in the research, fieldwork, or practical settings (see chapters 15: and 18: The student in a leadership, mentorship and supervision role and Human accommodation – laboratory/technical assistants in the sciences). Meanwhile, negative cultural attitudes toward disability on the part of peers and colleagues can translate to social isolation, feelings of impostor syndrome, stigma, and decreased mental health and well-being (see chapter 6: Disclosure in the sciences). Cultural perspectives on disability by parents or the student may result in a lack of willingness to disclose an accommodation need (see chapter 5: Key role of education providers in communication with students and service providers), or to engage effectively with faculty and educators (see chapter 11: Universal design for learning). Of course, cultural perspectives can also influence all stakeholders in the conversation simultaneously. Working through matters of cultural sensitivity around disability requires sensitivity, care, patience, and a willingness to understand the conversation through the other's lens. Returning to the fundamentals of the human-rights framework we defined in this chapter for STEM education is recommended as the starting point for a common dialogue.

Science as an international endeavor

More than most disciplines, the sciences are truly global in scope. Students may choose to go abroad for any or all levels of their postsecondary education. In the United States and Canada, approximately 50% of the postdoctoral cohort is international, with the greatest proportion of international trainees coming from China, India, and Western Europe. The flow of human capital at the faculty level is equally diverse. Internationally trained early career researchers may return to their native countries or communities. Nations looking to bootstrap their scientific research enterprise in a competitive global market may specifically target senior faculty candidates trained in Canada, the United States, the United Kingdom, or Western Europe or may make strategic partnerships with institutions in those countries—in recent years, Singapore and several countries in the Middle East have employed this strategy.

Even students, trainees, and faculty who choose not to go abroad (and, for students with disabilities, there may be logistical challenges or disability management considerations that may be limiting) can experience the globalization of science because of this international mobility. Faculty colleagues, student peers, and postdoctoral fellows from a diverse array of backgrounds may come to your campus, bringing scientific globalization to your doorstep.

Thus no longer is basic cultural competency a prerequisite only when going abroad—all science students, trainees, and faculty members must master this soft skill. In the context of disability, recognizing that your student, classmate, lab partner, collaborator, or faculty member comes from a different background than you, and perceives disability differently than you, is crucial.

Conclusion

In thinking through the barriers experienced by students and trainees with disabilities in the sciences, and in discussing the experiences of students with disabilities across the globe, we realized that while the specific details of circumstances and experiences might be different for different individuals, the underlying principles remain the same. Indeed, we realized that diversity of location, geography, and culture was no different in scope than diversity of discipline, disability, or level of education.

Finally, it is worth reinforcing the global lack of knowledge around STEM education and disability. Prevalence and participation data for persons with disabilities in STEM remain unclear, as do techniques for inclusion, training for educators, methodology, and best practices. The material in this book contains the first attempt at a holistic understanding of the barriers faced by students with disabilities in STEM, and it is our goal that this work begins to develop the conversation around participation in STEM and the body of knowledge around effective practices in STEM education and training for students with disabilities. The following chapters thus address barriers and their solutions from a global, cross-cultural, cross-disciplinary, and cross-disability perspective.

Part II

Barriers Faced by Students with Disabilities in the Sciences

Barriers faced by students with disabilities in science laboratory and practical space settings

Chapter Outline

Introduction   27
Occupational choice   28
Lack of professional development for service providers and educators   28
  Educators   28
  Service providers   29
Structural differences in student support systems between high school and postsecondary education   29
Student awareness of support systems in postsecondary   30
Student engagement with support systems in the educational setting   30
Lack of access to assistive technologies   31
Adapting mainstream technologies   31
Universal access to scientific materials   32
Availability of accessible formats   32
Logistical considerations in the lab and classroom   33
  Logistical considerations in the STEM laboratory, fieldwork and practicum environments   33
  Logistical considerations in the STEM classroom   34
Self-advocacy   34
Support network advocacy   35
Attitudes   35
Competing priorities in education   36
The "gatekeeper function"   37
  The challenge of misinformation   37
  The challenge of lack of information   38
  The challenge of inductive reasoning   38
Conclusion   38

For as long as I can remember, I wanted to be a scientist. Thinking back on it, I doubt my parents really knew what to do with that. Their experiences with children who had disabilities were limited to me, and I think they felt that they were in this all by themselves. It must've been hard – they wanted to be supportive, but didn't know how. They wanted to be realistic, but didn't want to crush my dreams. A difficult spot to be in – and the four year old me, announcing my

Creating a Culture of Accessibility in the Sciences. DOI: http://dx.doi.org/10.1016/B978-0-12-804037-9.00003-6

intention to the world, cared little for such nuances. The adult me, reflecting on it all, picks up on them easily.

The thing is, wanting to be a scientist and then making it happen are two very different things. There were so many people who said "You can't do that" or "You're the first person I know who wants to do that". The former I took as challenge. The latter, as flattery. I felt that those who said "I don't know how to help you do that" were the ones who came closest to the truth. In the end, I think the best of my teachers and professors ended up doing what my parents did – they didn't know what saying "yes" would entail, didn't want to say "no", and so did their best to figure things out with me. We all made it up as we went along. Not the perfect solution, perhaps, but the only one, really, that we could think of collectively.

Who knew there would have been so many challenges? So many different things to think about? I was captivated by books about science at an early age – almost before I could read and understand what I was reading, I suppose – but none of those books ever said what it was like to be a scientist! I thought it was all about thinking – and I was definitely good at that! Nothing talked about what it took to actually back up the thoughts, to prove or disprove a hypothesis. The logistics alone of working in a lab, when I figured them out, were staggering – the need for adaptive software, custom technologies, alternative formats. The physical and visual challenges of doing experiments, the need for human assistance. The need to be ultra careful about things like radioactive materials, dangerous chemicals, and biological hazards. It was almost too much – it was a good thing I had a few mentors who recognized my potential and wanted to help. Without that mentorship, I don't know where I would have ended up.

It was so different, too, coming to university for the first time – I had no involvement in my disability accommodations in high school. In fact, I didn't even know they were being negotiated around me until I was almost ready to graduate. I certainly didn't realize what I needed to know to put those accommodations into place – and when it was time to answer those questions in university, I was stuck. Getting used to that new environment, figuring out how to navigate it, and figuring out how to advocate for myself, were difficult. I needed to learn how to do these things, and as it happened, I had to learn on the fly. I think, if I had to do it over again, being better prepared for my undergrad – even for my master's – would have been the way to go. I finally learned my lesson with my doctorate – third time is the charm, right?

Even so, even with all the challenges, would I do it all over again if I could?

Absolutely – being a scientist is always what I wanted. For so long, no other career was even considered. Now that I am a scientist, I have come to learn two things: First, being a scientists means I can *do anything I want – the skills are totally transferable into so many career paths, so many walks of life. In many ways, that was the easy lesson to learn.*

The second – and much harder – lesson to learn is this: Being a scientist isn't a "thing" that we do, it's a "who" we are. It's a state of being, a state of personhood. It's not about what we do, it's about how we think. And none of that has anything to do with my disability. I think anyone can be a scientist – with the right character, the right mindset – and anyone can succeed at it, with the right nurturing environment.

Introduction

Much like the manner in which women have been historically discouraged from certain fields of "male-dominated" study, people with disabilities are discouraged from fields that are considered rigorous and requiring "precision." This perspective often extends to educators at the postsecondary level, who are in a position to foster the development of students with disabilities in lab-based science courses, curricula, and programs of study.

According to the National Science Foundation (NSF), barriers to success in science, technology, engineering, and mathematics (STEM) careers include:

- Diminished support systems after high school (students entering lab-based courses may not be aware of available supports in their university, or the supports simply may not be available);
- A total lack of support for students with disabilities taking science courses in high school is also possible, as neither teachers nor special education departments may know how to work with students in those areas;
- A lack of awareness of successful role models (students may not be aware that there are, indeed, successful scientists with disabilities from whom they can learn);
- A lack of access to technologies (students may not have access to the required assistive technology (AT) that would enable them to take part in lab activities);
- Poor self-advocacy skills on the part of students;
- In many cases it is not simply that the student can't manage to advocate (due to a lack of skills), but instead the student is put in a position where there are barriers to self-advocating (i.e., the attitude of educators: "if you can't do it just like everyone else, you can't handle this field," "I can find someone who doesn't require extra work" (in regard to making reasonable accommodations), "why are you even here if you have a disability?").
- Inadequate accommodations; and
- Low expectations from faculty— "she'll wander into something else when it gets too hard" (Hilliard, Dunston, McGlothin, & Duerstock, 2011).

In this chapter, we will review the major barriers faced by students with disabilities considering a journey into STEM fields and careers. As we discuss these barriers, we also highlight possible root causes and systemic issues, which we will return to and present potential solutions for later in the book.

Occupational choice

Students with disabilities are precluded from taking part in STEM, and thus, are denied the right to pursue this field, largely due to environmental and attitudinal barriers beyond the student's control. This lack of "choice" can be viewed as a form of occupational injustice, whereby students with disabilities are not only restricted from full inclusion and participation throughout their educational experience, but are not receiving accurate and appropriate information about STEM fields, which would thereby contribute to the student's ability to make an informed occupational choice. The ability to make daily autonomous occupational choices has been identified as a human right and positively linked to well-being. Those marginalized from choice are considered at risk of occupational injustice. Students with disabilities are one such group who may be marginalized from making such choices. Occupational choices do not appear to emanate from internal interests and abilities but rather from necessitated and mandated actions based on sociocultural expectations and erroneous assumptions concerning a student's abilities. This suggests that educators, professionals, families, and others who work with students with disabilities should pay closer attention to the social context that informs choices to improve the unecessary but all too common gap between students with disabilities who wish to pursue STEM and those actually entering the field.

Lack of professional development for service providers and educators

Educators

In contrast to teachers in the K-12 system, faculty and teaching assistants (TAs) on college and university campuses receive little in the way of formal instructor or teacher training to carry out their teaching duties, regardless of discipline or field of study. In the context of STEM fields, faculty and TAs must rely on their knowledge of the field and their own experiences as students to instruct their classes in the classroom and laboratory settings. Likewise, in the context of internships, co-ops, and the research setting, there is little formal mentor training and a reliance or bias on personal experiences in training students. For the purposes of our discussion within this book, we use the term "educator" to refer collectively to teachers in elementary and secondary schools, as well as postsecondary faculty, TAs, and mentors in STEM fields.

This reliance on personal experiences, coupled with the current low numbers of students with disabilities in STEM, leads to a lack of personal exposure to disability and accommodation issues on the part of educators involved in STEM training in postsecondary education. Students with disabilities may present as outside standard norms for teaching practices, and may need to learn in unconventional ways based on the interaction among their disabilities, discipline, and educator. Faculty and TAs then do not have personal experiences to build on in developing their instructional and mentorship skill sets to work effectively with students with disabilities in STEM. Educators are unable to rely effectively on their peer networks for support, as the low number of students with disabilities in STEM compound the personal experience challenge across the postsecondary system.

Educators may not be aware of resources on inclusive teaching practices and accessible course design that exist within institutional centers for teaching and learning excellence and across the teaching profession (Badenhorst, Moloney, Rosales, Dyer, and Run, 2015). Because of the reliance on personal experiences in developing teaching and mentorship methods in college and university, educators may not think to review these resources. Again, because of the relative lack of precedent, and therefore need within STEM, to work with students with disabilities who are interested in these fields, teaching resources may not encompass STEM-specific best practices that could be of benefit. Therefore, educators looking for coaching and resources on working with students with disabilities may be at a loss as to where to turn.

Service providers

Low numbers of students with disabilities in STEM also translates to a lack of professional experience accommodating STEM-related needs on the part of the disability service provider in postsecondary education. As service providers often rely on their professional networks to identify best practices and precedents to follow, this lack of professional experience is compounded across the postsecondary system as well. Service providers will often work with faculty and instructors to provide appropriate accommodation for students with disability; however, there may be a lack of familiarity between STEM faculty and service providers because of a lack of consistent need to engage together on disability issues. Furthermore, there may be a lack of knowledge on the part of the service provider about how and where to access resources, assistive devices, and other technologies that would better enable students with disabilities to participate in STEM labs and courses. Finally, service providers interact with senior academic administrators, who are faculty, in setting policy and practice guidelines for accommodation at the institutional level. The lack of experience and knowledge shared by both parties around working with students with disabilities in STEM can ultimately lead to a discouraging environment for students in pursuing their interests in science and technology. Such an environment often expresses itself in a paternalistic manner ("educator knows best" or "service provider knows best"), with decision-makers often making choices for students with disabilities without having a good understanding of their true abilities, limitations, or needs.

Structural differences in student support systems between high school and postsecondary education

In the kindergarten to grade 12 (K-12) education system, students with disabilities may be integrated within "mainstream" schools (even if placed within special education or special needs departments), or may be placed within segregated (sometimes even residential) schools, or may be homeschooled, or excluded from the education system altogether. Disability accommodation support can be formalized and structured in a specific, mandated, manner built around Individual Educational Plans and team-based case management (or their equivalents in other parts of the world), or may often be informal and/or ad hoc. Parents of students with disabilities drive disclosure of disability and accommodation needs, and a team of educators and specialists can be engaged to develop and implement a formal accommodation plan.

On college and university campuses, accommodation support is structured around specific accommodation offices working on behalf of the student, with information transfer around accommodation needs being more compartmentalized than in the K-12 system. Indeed, confidentially of disability and accommodation information is paramount, and only that which is necessary is provided. Furthermore, students drive disclosure in the postsecondary environment, and accommodations are determined by a specialist working with the student and on his or her behalf. Disclosure to educators is not required, although it may be encouraged and left up to students. Ideally students participate in defining and shaping their own accommodations in postsecondary education. Indeed, students are expected to be more independent and have a sense of their own accommodations.

Student awareness of support systems in postsecondary

The potential for the lack of student engagement in their accommodation planning in high school may translate to a lack of knowledge or awareness about the appropriate discussions to have in postsecondary education. Furthermore, the complexity of postsecondary institution structures, sources of funding, office mandates, etc., can be difficult for students to parse without intervention and guidance during the transition process. The student may know the most about his or her accommodation needs and abilities, and it is imperative for the student to fully participate in the accommodation process as a result. However, the student's role as "teacher" of educators and service providers in postsecondary education is a lot harder to execute if the student is not fully aware of the interface between his or her abilities and accommodations. Finally, students who are unaware of role models in their field, who themselves might be aware of supports in postsecondary, may choose to not pursue STEM as a consequence of this lack of knowledge. Thus if students are aware of and/or provided the appropriate supports and resources, it is more likely that they will pursue and be successful in STEM disciplines.

Student engagement with support systems in the educational setting

It is worth highlighting that some students, although aware of support systems in place in secondary or postsecondary school, may simply choose to not engage with them. Often, this is because of the fear of being told "you can't do this" and wanting to try something before being told that it cannot be done (and thus being denied the choice to engage with the field). The root of this barrier is the potential for paternalism in decision-making at the secondary and postsecondary levels that we identified earlier in this chapter.

Lack of access to assistive technologies

According to Mohler (2012), assistive technologies (AT) is "any item, piece of equipment or product system, whether acquired commercially off-the-shelf, modified or customized, that is used to increase, maintain or improve functional capabilities of individuals with disabilities." Examples of AT can include: screen readers, screen magnifiers, optical character recognition (OCR) software, closedcircuit televisions (CCTVs), speech-to-text engines, etc.

AT can bridge the communication and accessibility gaps for people with disabilities. Electronic communications provide options for independent access to information and resources. Unfortunately, many individuals with disabilities and people in their primary support systems are unaware of the tremendous contributions technological innovations can make to the lives of individuals with disabilities. Students with disabilities do not readily have access to the AT required to increase access in STEM course. This is largely due to funding constraints and a lack of knowledge on the part of educators and service providers as to how to acquire technology. If educators and other professionals are not aware of means to subsidize AT funding the options of STEM courses may not be pursued due to lack of awareness/access to potential technological solutions to barriers in courses. Additionally, students may not have received appropriate and frequent training in the use and application of AT, and therefore, cannot explain to others how AT can be used as an accommodation.

The compatibility of AT with other forms of IT and other forms of software in the STEM environment presents challenges if AT cannot be retrofitted to map onto laboratory machines and computers. Particularly in the context of new and fast-developing fields in STEM, such as genomics, it is important to evaluate AT software for compatibility with newly developed commercially available analysis software solutions and/or cloud-based computing tools. This lag in compatibility might then make it hard for the student, as well as instructors and mentors, to design effective workarounds and solutions to accommodation needs.

Adapting mainstream technologies

In addition to AT considerations, it is worthwhile noting that adaptation of off-the-shelf technologies may provide creative and low-cost solutions to student accommodation needs. In these cases, while the appropriate ATs may be available, it may be more cost-effective—and, indeed, may be more appropriate for the student—to custom-engineer solutions using established, off-the-shelf technologies that were not originally designed for accommodation purposes. For example, instead of designing and generating large-print labels for chemicals, a set of handheld magnifiers could be used instead for some students. Additional challenges are posed with these types of solutions, which require full collaboration among educators and service providers, and full participation of the student in the solution-development process. While educators may be more aware of the off-the-shelf technologies that might be helpful, the service providers and students will be more aware of how helpful a proposed solution might be.

Universal access to scientific materials

Those who wish to pursue STEM fields need access to publications in these fields, yet STEM publications are not always readily available in alternate formats. Making them available in an accessible electronic format is desirable, but some barriers still exist—for example, in making mathematics and scientific symbols and graphic images accessible to those who are blind. Universal access to publications will require the creation of new products, as well as promotion of the use of existing methods. Webmasters also need to apply standards, such as those used by the US government, in order to take steps toward making their resources accessible to individuals with disabilities, including those who are blind and use text-to-speech technology.

Availability of accessible formats

Accessible formats consist of different means of producing materials such that the learner can intake the same materials presented in a different way. For example, a book can be produced in many different forms (audio, Braille, e-Text, large print, tactile). One major barrier to assessing information for learners is the conversion of symbology and graphics into formats accessible to those with print disabilities. Often, science and math concepts include the presentation of equations, graphics, charts, and other visual representations that are difficult to describe by text alone. Another informational barrier is the technical nature of the fields and the appropriate "translations" for highly complex technical scientific, engineering, and mathematical terminology. Knowledge of and cost associated with the method of conversion is also a key issue—e.g., are there local agencies capable of performing the conversion.

Given the multiple formats that text and other course materials can be presented in, copyright can become an issue for electronic conversion. Not all course material may be copied (intellectual property issues on the part of the faculty or instructor can pose a significant barrier).

Time for material format conversion in the context of course work also poses a significant challenge. A lag of 6 months is inappropriate for a 13-week course. How should the student receive instruction in accessible formats when the instructors may not be trained in this manner of information delivery and instruction? Accessible formats may also be required on the fly in the lab setting for print-disabled students (e.g., instrumentation, labels of chemicals, and equipment).

Logistical considerations in the lab and classroom

Often the volume of logistical considerations that need to be considered to ensure full participation of a student with a disability in a STEM laboratory (both instructional and research settings) or classroom can be overwhelming. Careful preparation and advance planning on the part of the educators, service providers, and students is essential in minimizing challenges as the student becomes integrated into the lab and classroom environments, but knowing which questions to ask and where to turn to seek answers can be daunting. Furthermore, the interaction between and among logistical considerations greatly increases the complexity of the discussion as well as the potential solutions. Taken together, the weight of these perceived unknowns can form a significant barrier to students with disabilities participating in STEM laboratory and classroom environments. What follows is a brief summary of some of the logistical considerations that need to be taken into account in these two settings.

Logistical considerations in the STEM laboratory, fieldwork and practicum environments

Barriers faced in the context of integrating students with disabilities in the STEM lab environment, in the context of both research labs and instructional settings, include:

- The need to install adaptive software on lab computers/computer-aided instruments, and the need to ensure compatibility between software tools.
- Procurement can serve as a logistical challenge in this context as well, as there is a need to ensure that lab computers are accessible and that sufficient and appropriate licenses for assistive software have been obtained; this might require interaction with institutional and departmental IT and purchasing divisions, who may be at best unfamiliar with accessibility considerations.
- Lack of alternative formatting for lab standard materials—e.g., safety sheets, signage, chemical labels, etc. These documents and labels are often printed in small fonts.
- Lack of modified laboratory equipment for accessible data collection (e.g., talking pH meters).
- Physical accessibility of the lab—standard specifications for fume hoods, sinks, aisles, benches, etc., do not permit ease of access for persons with physical disabilities.
- Safety and workplace hazardous materials information safety training and preparation can also pose challenges in the context of a student's disability(ies)—in particular, whether the student is likely to be in a position to handle materials that might prove more difficult or dangerous to use given his or her disability.
- Working with biohazards and radioactive materials introduces heightened safety concerns, and may require considering different and more creative designs to experiments the student needs to conduct in the laboratory environment.
- The potential for lack of manual dexterity for experiments and techniques involving fine motor skills can also be a concern, and increases the challenge of students needing assistance at the bench to perform the experiments.
- The buddy system—or provision of a laboratory technical assistant—can be potentially disruptive to the lab team dynamic and requires flexibility to implement effectively.
- Finally, the educator may be unsure of how to measure the technical competencies of the student in the context of the above logistical considerations—this is particularly concerning when thinking through the essential requirements of the course, program, or discipline, and is the subject of a chapter later in this book.

Of note, many of these considerations also apply in the context of working in the field and in practicum environments in STEM and related disciplines.

Logistical considerations in the STEM classroom

Barriers faced in the context of integrating students with disabilities in the STEM classroom environment include:

- Format of course materials.
- Format of course assessments.
- Accessibility of course-content delivery.
- Physical accessibility of classroom environments.
- Interface between the student's technology and the classroom environment.
- Finally, as with measurement of competencies in the STEM lab environment, the educator may be unsure of how to measure the technical competencies of the student in the context of the above logistical considerations in the classroom—this is particularly concerning when thinking through the essential requirements of the course, program, or discipline, and is the subject of a chapter later in this book.

Self-advocacy

Self-advocacy is the ability to communicate and work toward the implementation of one's needs, accommodations, and requests for resources to facilitate full participation in the community.

In order for students with disabilities to be successful, it is necessary to develop strong self-advocacy skills (Burgstahler, 1994; Street et al., 2012). A systemic barrier to developing strong self-advocacy skills is built into the education system. High school and undergraduate education too often provide a structured environment whereby educators and service providers guide much, if not all, of the student's advocacy and accommodations. When students are not permitted the opportunity to practice self-advocacy in these settings, students do not learn how to ask for and implement their accommodations. When students permit the service provider to advocate on their behalf, this can dilute student influence over the final outcome. It is critical to start working with students early on to develop self-advocacy skills, For example, communicating with teachers about needs, communicating with peers, etc.

Students with disabilities are rarely encouraged by educators, service providers, peers, and family to prepare for STEM fields. Since they do not consider a career in STEM an achievable goal, they do not take the courses necessary to prepare for postsecondary studies in these areas (Street et al., 2012). Educators often lack an understanding of the content and requirements of STEM programs in higher education and of the technology and other resources that make it possible for students with disabilities to pursue these fields.

Students with disabilities seldom have access to mentors with or without disabilities in STEM fields who are successful in careers they might otherwise have thought impossible for themselves.

Support network advocacy

Support network advocacy is advocacy on behalf of the student that is conducted by parents, peers, service providers, educators, and other professionals. The advocacy that takes place by the support network is not under the direct control of the student; indeed, the student may not be fully aware of advocacy efforts undertaken on his or her behalf by the support network. These efforts may influence where and how the student can be successful if messages and the student's needs or choices for course of study get lost in translation. For example, several people who have differing backgrounds, experiences, philosophies, and perspectives on the student's abilities and skills may be advocating on the student's behalf. Without ensuring that these perspectives are harmonized, individuals may communicate differently about the student's needs to the same individual or office, leading to the potential for confusion and improper or inappropriate accommodation.

If advocacy is done improperly, students may not have opportunities to pursue STEM disciplines, or to select fields of study that best meet their skillsets. The lack of required STEM accommodations may be influenced by the lack of correct knowledge and skills surrounding how to advocate for the necessary technologies that would permit a student to succeed in STEM. Competing perspectives and priorities of individuals involved in support network advocacy can give rise to the same result. Additionally, in the case of support network advocates within the education system, their ability to bring funding and other resources to bear in accommodation discussions may impact the nature of those conversations. Strong advocacy is influenced by adequate preparation, research, and background knowledge on the part of the student's advocates.

It is therefore critical that there is coordination and collaboration when advocating on behalf of a student. A lack of coordination leads to mixed messaging and inaccurate representation of the student's needs. Inadequate collaboration leads to the potential for unclear and muddled communication and inappropriate accommodations. This can be exacerbated by "helicoptering"—the overparticipation in advocacy on the student's behalf by someone other than the student (e.g., parent, educator, service provider), with the resultant overexpression of the perspectives of the advocate rather than the student's own needs, opinions, or perspectives.

Attitudes

Students can be discouraged from an early age and throughout their educational journey from pursuing STEM-related disciplines by negative attitudes expressed by family, support staff, and educators. Negative attitudes may come from comfort levels that stem from singular ways of teaching and learning in STEM. Negative attitudes may evolve from how educators are trained, as well as from an inability to shift their way of "doing" in order to be open and interested in new ways of thinking, teaching, and doing.

More training is required around STEM and disability, which will build a culture of inclusion and may lead to increased numbers of students going into STEM fields. Educators, guidance and career counselors and support staff, in particular, are not discussing STEM with their students. If a discussion does take place, it is not often a positive one, and not taking place with STEM in mind as a possible postsecondary education path followed by a potential career option. Through streaming into non-STEM related courses, options are not being presented (constituting a negative attitude or perspective by omission). The notion that "it's never been done before—therefore it can't be done" is prevalent in the mainstream STEM educational setting.

Students are also actively excluded from STEM when they are deliberately coached in a direction away from the sciences, because the educators believe that STEM is not a viable option. Conversely, students are passively excluded from STEM when a discussion is never had with the student regarding the possibility of STEM training and career paths, and the student is never presented with all of his or her options.

Attitudes may not be "negative" but may result from individuals not having a solid understanding of the many possibilities for pursuing various careers within STEM. It is often simply that families and educators do not know how and where to acquire resources, or that they do not have access to resources that may aid students. Educators and family may not be aware that there are indeed successful individuals with disabilities pursuing careers in STEM, and that these individuals can also act as mentors for the student and educators.

As a result of lack of knowledge surrounding resources, mentors in the field, and appropriate accommodations, students may be steered away from courses in STEM, thereby not enabling them to pursue STEM at a higher education or career level. Because of these attitudes, students may not tend to believe it is possible to have a career in STEM. Proper mentorship (for educators and service providers) is as important as mentorship for the students, as appropriate guidance can lead to an amelioration or negation of these attitudes (Jekielek, Moore & Hair, 2002).

An additional double standard exists among educators, parents, service providers, and mentors, in that they may actively steer students away from STEM education because they think it will be "too hard" or the student will be presented with a higher risk of failure. Every student has the potential to fail at an attempt at a STEM course or career, but students with disabilities are seen as "too fragile" for such life lessons and experiences such as failure. What might be viewed as a valuable experience for a nondisabled student would be seen as "too devastating" for a student with a disability.

Competing priorities in education

One additional factor—social interpretations of individuals' functional limitations due to disability—may contribute to attitudinal barriers on the part of educators, parents, service providers, and others within the student's support system. As they proceed through primary and secondary education settings, whether they are in integrated classes, or in special education classes, or in segregated and/or residential schooling, students may be in situations where instruction in math and science is deemphasized in favor of independent living or other skill-based instruction (e.g., Braille classes for students with visual impairments). This altered emphasis in education may be due to fundamental biases around the ability of students to fully participate in—or, even more basic, to require instruction around—STEM-related fields. For example, basic math competencies are required today to balance a checkbook, understand student loans and interest rates, and manage long-term investments and mortgages. However, an educator's perception may distill down to "The student can't afford a mortgage on disability assistance anyway, so why bother teaching the skillset?"

Not only do negative attitudes have an impact on whether a student will enter into STEM disciplines, but negative attitudes or attitudes that question the student's ability in STEM contribute to the fact that we, as persons with disabilities (students, employees, trainees), have to prove our abilities more so than our peers (Mohler, 2012). Given the incorrect, but all too common, assumptions concerning the academic abilities of students with disabilities, it is not surprising that we as students are not only proving how we can perform the same or better than our peers, but are disproving a host of myths and misconceptions regarding our fit in nontraditional fields. Of course, all students, regardless of ability, have to continually prove their skill level, but this is exacerbated when a student has a disability and his or her abilities are consistently being questioned, especially if the student is in a STEM-related field. Students who are consistently running up against negative or questioning attitudes have to prove they "belong" in the field. Too often, our need to prove our capabilities and fit for a given field results from educators and other professionals informing us we "can't do it" or "I don't care how much training you have, my perception says you can't do this job." It is therefore up to us, as students with disabilities, to not only prove that we belong in these uncharted waters, but also that these spaces ought to be made more inclusive and accessible for future students with disabilities.

The "gatekeeper function"

Educators and service providers, when faced with the challenge of accommodating and working with students with disabilities in STEM fields, may often believe based on their experience that it is not possible for the student to succeed in the discipline or program of study. This belief contributes to a reluctance on the part of the educator or service provider to work effectively with the student, and might even lead to active resistance to the student's participation in the course, program, or discipline. We call this the "gatekeeper function."

Itself a barrier to student success, the gatekeeper function arises from the attitude that the student ultimately is incapable, or is not fit, for the course, program, or discipline. This belief may arise from any of three key notions as follows.

The challenge of misinformation

An educator may be aware, secondhand, of other students with disabilities, who have attempted to succeed in a given field of science. These students may have failed in their efforts due to poor academic standing and/or inappropriate accommodations. However, the educator may not be fully aware of the complexities and needs of that student and therefore has drawn the wrong conclusions about the student's ability. The educator thus makes a false assumption based on incorrect information. This can easily occur, as disability disclosure, accommodation need, and the accommodations themselves, are often protected behind layers of privacy—further, the particulars of the case can get warped with time and distance, as the information spreads through the educator's network to him or her.

The challenge of lack of information

A disability services officer may be asked to work with a student with a disability in the context of a STEM field in order to provide the appropriate accommodations. Since students with disabilities have historically been a rare breed in STEM, the service provider has no contacts in his or her network who can assist in determining the appropriate accommodations. Furthermore, the service provider may not know the right approach. This sets up a lack of knowledge regarding how students in STEM fields can succeed with appropriate accommodation. Thus the service provider falls victim to the misperception that not knowing about accommodations translates to there being no accommodations and thus the student is not able to succeed in his or her chosen field.

The challenge of inductive reasoning

The educator and service provider meet a student with a disability intending on pursuing a career in STEM. The educator is aware of another student he or she has worked with who succeeded with a defined set of accommodations, and together the educator and service provider develop an accommodation plan for the student based on this prior experience. However, this approach does not take into account the specific needs of this student, and fails because the accommodations were inappropriate. Here, the educator and service provider fall prey to the trap of inductive reasoning, and make the mistake of attempting to take advantage of their previous knowledge in an inappropriate setting. This has the consequence of doing more harm than good, as the student ends up with inappropriate accommodations and is set up for failure.

Conclusion

Students with disabilities are not well represented in STEM disciplines. Barriers such as lack of mentors in the field, lack of knowledge regarding how and where to acquire AT, a knowledge gap surrounding how to instruct a student with a disability, and few opportunities for students to pursue STEM at a young age all play a role in limiting access to STEM. Those in the educational setting have few opportunities to instruct students with disabilities in STEM disciplines and therefore do not develop the necessary comfort level, requisite skills and tools that would enable a student to be successful. It is critical that students who wish to pursue STEM-related courses develop strong self-advocacy skills and have a strong sense of their disability-related accommodations. Doing so will allow them to more accurately speak to their educational needs and how to accommodate them. It is also strongly recommended that students have access to a mentor or champion that can guide them through the process of navigating the myriad barriers in STEM.

As we will discuss in subsequent chapters, it is of great importance for educators, family, and service providers to not make assumptions based on past experience, or lack thereof, and instead to work closely with students to understand their needs and the interface between the course, program, discipline, and the specific accommodation needs. Such an understanding, and a willingness to work with the student and keep an open mind, will foster success.

Student perspectives on disability—impact on education, career path, and accommodation

Chapter Outline

Introduction   41
"What could they do for me?"   42
"Ignorance is bliss"   43
"Raising the bar"   45
"There is always a Way, it's just a matter of finding it"   47
"Easier said than done"   48
"Did my opinion matter?"   49
"Not being good enough"   49
"Knowledge is preparedness"   52
"Disability is not something to be ashamed of"   53
"What could have been?"   53
Conclusion   54

Introduction

In chapter 3, Barriers faced by students with disabilities in science laboratory and practical space settings, we highlighted the array and complexity of barriers that are faced by students with disabilities seeking to pursue education and training in the sciences. In this chapter, we present the stories of individuals who have experienced these barriers, and, where possible in the context of their circumstances, identified and implemented solutions. In this chapter, the narrative of the student voice and the student experience takes center stage, and we invite you, the reader, to distill the essence of learning for yourself. In our experience, the lived experience is a powerful communication tool, and here we use it to highlight the importance of understanding the context of a student's journey and how that context and journey will ultimately impact a student's attitude and willingness to engage in dialogue around disability and accommodation—as well as how that context and journey influences a student's perseverance in his or her science education and training.

Creating a Culture of Accessibility in the Sciences. DOI: http://dx.doi.org/10.1016/B978-0-12-804037-9.00004-8

"What could they do for me?"

As an undergraduate in the sciences, I was navigating through new waters with a secret that I felt must be concealed to the best of my ability. For me, growing up with a disability had been something to be discrete about. I actually felt a sense of accomplishment, as if I had done something "good," when I was in a situation where I was "passing" as nondisabled. To accept accommodations of any type was to admit to weakness, or to "play the blind card," so I would push myself to go unnoticed for as long as I could. Teachers would be amazed and praise my abilities, my parents would be pleased, and I would be left with eye-strain migraines, hours of extra work, and an impending sense of failure. I actually did reach the limits of my abilities in my sophomore year of college. I was at a cross-roads—I would either have to accept accommodations, or start to have my grades affected by my fear of admitting that I needed a more evenly leveled playing field.

Despite finally coming to a place where I had open communication with my university's disability services office, I was still uncomfortable discussing accommodations with professors. Aside from the trouble explaining that just using a classmate's notes was not enough (since I cannot read most people's handwriting to save my life), I was afraid of describing the obstacles I encountered in a given course and having the professor tell me they didn't know how they could accommodate and that I shouldn't take the class. Essentially, I was afraid of being told that I couldn't before I even tried. This has always been my fear.

Growing up, I was raised with the expectation that I would accept the decision of an authority figure without question, and I was left feeling very vulnerable. When I entered college with this mindset, I lived in perpetual anxiety that I would be told my goals were unobtainable and that I needed to change the course of my life. Knowing that I could find a way to navigate obstacles differently from my classmates, I actually avoided potential mentors out of fear of the gravity of their opinions and inexperience with working with people with disabilities. I had never met another legally blind scientist before the summer of 2014, and the thought of finding a mentor with a disability had never crossed my mind before this time because I was under the impression that no one could be openly disabled and a scientist.

The only drive that was stronger than my anxiety was my attraction to the field of biology. My earliest memory of wanting to be a biologist was when I was 5-years-old and was mesmerized by the star of the Warner Brother's movie "Free Willy." However, my parents will testify that my early obsessions with fishing shows and salamanders were the first real hint of my future aspirations. Not only was I fascinated by animals, but I also wanted to protect them. For my entire childhood, grade-school teachers always seemed to try to dissuade me from the sciences because they were either perceived to be "too hard" or it was implied that I could not handle the possible failures or hard lessons that accompany scientific study. It has always been a wonder of mine as to why it would be acceptable for a nondisabled person to experience struggle in the pursuit of greatness, but it would not be acceptable for someone with a disability. After all, could I not learn from my failures as well?

Once in college, I have to say that the disabilities services office were quite happy to begin working with me. The office housed what must have been considered an antique braille embossing machine, even before the turn of the new millennium, in addition to other largely unused low-vision tools such as enormously heavy black and white closed-circuit televisions and weak hand magnifiers. The office worked predominantly with students with learning disabilities, and a student with a physical disability was a little exciting to them. I knew this excitement all too well. Throughout my K-12 years I had always had to decline accommodations that were either excessive (i.e., braille, audiobooks, extremely large print) or accommodations that were insufficient (weak hand magnifiers or book stands).

I can clearly recall being offered both 2 X hand magnifiers and a ridiculously large and loud camera with goggles that was supposed to be a "hands-free vision aid." The only problem with the magnifiers was that they were far too weak to be of any real help, and the camera was so loud that I would not have been able to hear the teacher over its roar. There never seemed to be any middle ground until I finally received a 7 X magnifier during my senior year of secondary school. At the undergraduate level, I tried to go as unnoticed as possible and, aside from taking my exams at the disability services office so that I could use the time and a half accommodation as well as using a computer to write essays, I did not work with a disability counselor. Honestly, I wasn't sure what they could do for me. I was disabled, but my needs and the perceived needs of a stereotypical blind person were not in sync.

"Ignorance is bliss"

Growing up in the Caribbean, as a partially sighted child, was an interesting experience—I had no appreciation at the time that my childhood was any different from that of my siblings, or of any other child. A significant reason for that was that there were, in the 1980s (indeed, still to this day), no significant supports for blind and partially sighted children, nor were there support mechanisms for their parents and families. Indeed, the only resources we could draw on, first in Guyana, and subsequently in Jamaica and Barbados, were ophthalmologists—none of whom were pediatric specialists.

Reflecting on that experience, I recognize that my parents had to operate in near-impossible circumstances, with negligible levels of support. There was no one qualified to tell them what their child could do ... and, perhaps more importantly, there was no one qualified to tell them what their child could NOT do. Indeed, for the first 10 years of my life, my parents had to invent their own ways of dealing with my care, my schooling, my social life, and my interests.

When we emigrated to Canada, my parents encountered a support network in the form of the Canadian National Institute for the Blind (CNIB), and the special education department of my high school. Thinking about their interactions with those support systems, I see that my parents weren't fully comfortable with them, or the advice they were offering. Certainly, that advice was founded on experience, and may have worked in the majority of cases—my parents' thinking was that it didn't hold with their child, and the situations they had evolved and become comfortable with. So, even after coming to this country, my parents continued to develop their own ways of coping and engagement.

As a result of these, and other, choices, my parents gave me three intangible, but very powerful, gifts:

First, the ability to learn in my own space and pursue my own desires and dreams when it came to my education and career. For as long as I can remember, I have always wanted to be a scientist. I honestly don't know if my parents knew what to do with that desire, but I am quite certain that they knew what NOT to do with it. In not ever saying to me that this was an unattainable dream, in fostering as best they could my interest and desire to learn, and in not accessing advice that would not have encouraged my academic and career path, they gave me the space to learn about something that interested me, without the constriction of preconceived notions about what I could or could not do.

Second, the social and transferable tools I would need to execute my chosen path and gain the ability to make a difference in society. Without the benefit of research or the growing literature on the benefits of socialization to blind and partially sighted youth, my parents recognized the important and fundamental requirement that I learn organizational, communication, and leadership skills. To achieve this goal, my parents fostered my community and volunteer engagement at a young age—and after a 21-year career as a volunteer and successful leader in the communities I serve (conducted in parallel with my academic and scientific career), I appreciate this gift all the more, because I see its value in all aspects of what I do.

Third, and most importantly, by their example, the courage and moral fiber to hold fast to my dreams and convictions in the face of conventional thought. Specialists in the Caribbean had discouraged any significant investment on my parents' part in my development and education—even today, children with disabilities have little in the way of positive supports in that part of the world. In parallel, as I advanced through my education, teachers, instructors, and faculty were at various points doubtful of—or outright hostile toward—my ability to succeed as a partially sighted practicing scientist. To be sure, their doubt and hostility were founded on their experiences—and they most certainly had no experience with a blind aspiring geneticist. Today, thanks to perseverance, skill, opportunity, and hard work, I am North America's only blind geneticist and cancer researcher.

I think many might argue against my parents' choices—in fact, it's easy to argue against my parents' choices. I didn't get orientation and mobility training until I was 18. I never received independent living skills training. Certainly, those choices were not easy, cheap, or safe. Certainly, there was no safety net for a long period of time. Certainly, when a support system became available, it would have been easy to accede to their greater expertise. My parents didn't know what they didn't know, and, in their own way, were striking out into the unknown. In hindsight, there were probably a thousand ways of doing things differently, or better. We don't have the luxury of living in hindsight, though—we must make decisions as life puts them in front of us.

I have often wondered how life would have been different had I and my parents had full access to child and youth services supports—if, despite the expertise and support systems available, some of the opportunities and choices I had would have not materialized in that different reality.

"Raising the bar"

Growing up, my parents' approach was a balance of safety and security blended with an openness to me trying new things; as a younger child and young teen I was supervised more closely to ensure I wouldn't injure myself or put myself in a dangerous situation—but as I grew older and demonstrated a great deal of capability—that level of supervision decreased. My parents' approach was also comfort-based—if I asked for help or indicated I needed it, it was provided; otherwise, I would function well on my own.

My parents and educators never put my disability at the forefront of my inclusion or experience with anything—it was always a critical factor to consider, but never a deciding factor. I started downhill skiing at age 3, piano at 5, track and field at 10, and horseback riding at 10—I was encouraged to do these activities when I demonstrated an interest, and my vision was factored into the approach of me doing those activities, but it was always factored in versus serving as a driving or deciding factor—I was almost always able to be me first and me with vision impairment/disability second—and this attitude (driven by my parents and educators) has carried over into my own mentality and self-confidence about living life with disability versus letting disability drive the course of my life.

My parents' and educators' attitudes also had a great impact on the course of my education—my parents enrolled me in a private school in order to receive more attention in classrooms, smaller class sizes, and more specialized education options—they wanted to give me opportunities in life and saw this as a fundamental step toward me succeeding in an educational environment. A nurturing school environment (like the one I attended) was a good fit and ultimately contributed to a lot of the success I have achieved today—an important thing to note would be that my parents didn't just consider my disability when making a decision about my education—they also considered my own personality and how that meshed with or interacted with my disability; basically, they helped spur my childhood in a positive direction academically and personally by looking at me as a whole person—not just as a child with a disability.

I was always encouraged to participate as much as any other child or youth in school—in the classroom and in extracurricular activities. If my disability prevented me from directly participating in a class activity such as science lab—my teachers would allow me to do a one-on-one session with them outside of class time so that I could learn the material firsthand just like my peers. Being able to *do* things and not just watch them or listen to them is an attitude and understanding that my parents and educators had and it helped me greatly. Building an understanding of my preferred learning style as well as my needs based on my disability was also critical for me.

I was also encouraged to interact with peers and develop the social side of education and team building. I acted in plays, did student council, was on sports teams, and did art just like any other kid or teen—because I was interested and my disability would be worked around in order for me to join in. Joining in is so important, and again, it's an attitude that parents and educators can help drive forward.

Having accommodations and the framing of those accommodations is also critical to your own developed attitudes toward your disability. I have always been encouraged to look at my disability as something I live with, but not as something that holds me back. Special treatment was not something encouraged by my family or educators, but accommodations that evened the playing field and allowed full access to education were certainly a focus point.

I was always made to feel capable and able to learn and grow—and this is thanks to the forward-thinking and inclusive attitudes of my parents and educators. I think the perspective adults and role models have plays a huge role in how you come to see yourself. Yes, you can overcome damaging attitudes, being siloed, and being treated as less-than, but you can also overcome so much when you are challenged and treated like other kids your age—it builds a different kind of resilience, I think—knowing that you live with something that sometimes poses a struggle or barrier, but that others believe you are capable and make you want to rise to that challenge without settling into the easier trap of bowing out, leaning on the disability as a reason to not do things or not achieve as much.

My educators and parents created the space for me to have a well-rounded and "normalized" childhood if you will, though I hesitate to use the term normal. I felt like a kid and got to be a kid, I got to experience my teen years and grow into adulthood feeling a part of things versus apart, feeling like a person not a disability. This inclusion is perhaps the most important attitude that has shaped me, that and the challenge to rise higher than what is difficult and what feels hard one day—knowing it is making you strong and capable of tackling your future.

My experience as a student and young person moving through high school and into higher education became much more about raising the bar even higher. While educators and my parents wanted me to succeed, I found that rather than just meet the benchmark for success, they motivated me to excel and exceed that standard. Academically I received top marks and part of this was due to extra hours of tutoring in subjects like math and science where so much of the teaching is live and visual—but a larger part was the narrative I adopted under the influence of my parents and educators—that I can do just as much, if not more, than any other able-bodied student.

I think it is important for young people with disabilities to not be overly coddled, sheltered, or restricted in what they are told they can achieve in their education, professionals, and their lives. I was never told to stop, never told I'd reached the limit, I was only ever encouraged to push higher and do even better. This influence from adults I respected has continued to shape me as a working professional to this day.

I often approach a project, a problem, or an idea by looking beyond what the basic solution is—by thinking long-term about what the best possible way to move that concept forward into a successful outcome. This mentality is no doubt a byproduct of the attitude that was instilled in me as a young person in the education system—to always reach higher.

Looking back, this wasn't always easy, and sometimes as a person who is differently abled, reaching the baseline of success already feels like you've successfully run a marathon. Nevertheless, I'm glad no one told me I was done, that I could quit, or that I had completed what I needed to do. The door was always open for me to do more—and now in my line of philanthropic and charitable work, I am so very grateful for the lasting impact that my childhood experience has led to.

"There is always a Way, it's just a matter of finding it"

I was fortunate to have grown up by a provincial park. My playtime was spent outdoors; this likely sparked my fascination with nature. I was born with low vision, which decreased over time and by my 20s I had little light perception. In high school, I did well at first but began to struggle in my grade 11 courses, particularly in chemistry and calculus. My parents and a supportive teacher recognized that traditional blackboard teaching posed barriers, but it was challenging to find suitable alternatives.

In high school my career interest was in veterinarian medicine—I loved animals. The adults in my life, while supportive, realized that it was unlikely that I'd become a vet; my struggles in math and sciences supported this notion. A disconnect continued between my science-related career interests and my academic performance in high school. I was encouraged to explore the social sciences; I was informed that most blind adults were either social workers or massage therapists. I decided to go to college in a social services worker program. By this point, I had very little confidence in my academic abilities, and I chose this program specifically because it had no math or science requirements.

College was eye opening … suddenly I found myself as a top honors student. Following college I struggled for several years to find employment. After working in the disability field for 6 years I realized it was time to take control of my career. I explored options and discovered environmental studies and felt this was the right undergrad program for me. I loved it and excelled as an honors student. It helped me overcome some of those deep beliefs that I couldn't do science.

After thoroughly having enjoyed my undergrad, I was eager to pursue a master's. I chose a master's in forest conservation as my ecological fascination centers around trees, likely a result of my childhood environment. I chose a course-based professional master's. I longed to do research too, but I was conflicted about which path would be best; I was not confident that I could do good research as I still did not have a solid foundation in science and statistics.

As a blind student, there were numerous challenges in taking my course-based program. Most courses presented complex information visually; I hired tutors to provide descriptive assistance with funding through the accessibility services office. This funding also enabled me to hire assistants for my fieldwork courses to assist with data collection. It my experience when I requested accommodations such as course substitutions or alternative exam methods the faculty response was always no. The very nature of professional master's programs is inherent with barriers. They welcome diversity as long as it doesn't require doing things differently. Unfortunately, enormous amounts of my energy have been spent on trying to fit the mold rather than on pursuing my academic goals.

Finding my way to science has been a winding journey. I continue to move in the direction of ecology, perhaps that connection could have been made during high school, if only the accessibility issues in science and math hadn't been so distracting.

"Easier said than done"

For as long as I can recall, my parents have encouraged me to be active, engaged, and to take full advantage of opportunities both in my education and outside the classroom. Growing up, my disability was seen as but one of my many characteristics, and not something that should limit me in any capacity. From a young age, I have memories of my parents adapting everyday tasks (such as creating tactile maps to aid me in navigating my community) so I could be as independent as possible. Education was always a core value in my family. Both my parents understood the important role education plays in anyone's success in this world, but especially, the critical need it fulfills for someone who has restricted vision. So, when I commented that I hoped to pursue university education, steps were put in place to ensure I'd have the best chance to do so. I remember my parents asking me "so, what do you want to pursue? What are you good at? What are you interested in?" There was never conversation about how I'd need to select a specific career because of my vision restriction. Preparing for the road ahead not only included a focus on my academics, but on my volunteer, community, and leadership opportunities. After all, what university entrance committee doesn't like a well-rounded student?

However, I say all this knowing that it was not without their fair share of obstacles that my parents aided me in pursuing my quest for higher education. Professionals charged with educating parents and their children around disability and adapting to living with a disability were not always supportive of a family who wanted their blind daughter to have a "normal" upbringing. Like any other young adult, my parents had realistic expectations of me. I was expected to volunteer, obtain a part-time job, and maintain a strong GPA in school. One of the strongest life lessons I had instilled in me was that I was no different than any other student and that I should not expect to be treated as such. I think this has, in many ways, shaped me to become a strong self-advocate today not only for myself but for others within the disability community.

Too often, we, as students with disabilities, are streamlined into careers that best suit our disability, not our abilities and diverse abilities. For me growing up, my parents encouraged me to take what I was passionate about and turn that into a study or career interest. However, this was easier said than actually done. I recall being told in high school while I was in my last year of secondary, "you'll never go to postsecondary and certainly not graduate school." At this point in my educational journey, I was not presented with multiple options for reading and retaining information, and I was reading/receiving information in Braille, which was very slow for me. Once at postsecondary, I was properly assessed and found that audio enabled me to read/process much more efficiently and accurately. Had I known this earlier on in my educational career, I could have been better prepared for university.

What I think has most shaped me through my early experiences, both familial and instructional, is the huge role that others' attitudes, preconceived notions, and experiences have played in my life, and how they can shape and ultimately guide our educational and career path.

"Did my opinion matter?"

I attended mainstream schools throughout my childhood education because my parents believed that this was the best way for me to learn. As a child, I was quiet and shy, which was made to feel like a weakness by some teachers and support staff. These attitudes impacted how I perceived myself, as well as my self-confidence. It took the continued reassurance of my parents for me to accept myself in spite of what was being suggested about me. When I reached high school, the support staff were very patient and persistent in ensuring that I felt comfortable expressing myself and letting them know when something was wrong. However, while I become more at ease with sharing my opinions, I was not always convinced that they were valued in a broader sense because my views would sometimes be sought only to then be ignored, making the process seem superficial. Additionally, while I knew when things felt wrong, I often struggled to pinpoint the source of my unrest. For example, early on in high school, there was an expectation that because one of my peers with a disability had chosen to complete a physical education qualification that I must do the same. My preferences and circumstances were secondary to the school's desire for me to conform to the norm.

"Not being good enough"

First, a favorite subject—science! It's the easiest to start with.

I don't actually recall studying much science until 6th grade: I remember rocks and minerals; we actually did some "experiments" and by that I simply mean hands-on activities. it was probably at this point that we first heard about the scientific method. In grades 7 and 8, I found out I was "good" at science when I placed in the top three of my grade in the school's science fair—and though I was "neat and tidy" on my project, I simply found a somewhat interesting but to the point quick experiment from a book about science fair projects and whipped the stupid (shows my attitude toward it) project up at the last minute. I remember being extremely surprised my project was noteworthy at all—I'd grown accustomed to believing I was "stupid," and this feeling probably didn't leave me fully until the first or second year of undergrad—after lots of academic success, and more to the point, having met a group of hard-of-hearing students my own age over a Thanksgiving weekend leadership retreat in my second year of university (at that point, I was in a biology/concurrent education program).

In high school, I hid my hearing loss. Since most people don't recognize a hard-of-hearing person, and I never wore hearing aids or had any accommodations (i.e., assistive listening devices) in secondary school, this actually wasn't so hard. My friends just thought I was "daydreaming" or "out of it otherwise" when I didn't hear them. There was the "dancing in the hallways" when I tried to position them on my hearing (right) side but that was about it. A few of my teachers knew—mostly from my parents telling them. I certainly didn't say anything to any teacher until grade 12 or 13. However, I indicated to a grade-10 teacher my preference to sit near the board because I had trouble seeing (it wasn't untrue, I got glasses a year later and my parents were shocked).

Early on in my education I had a lot of ear infections, and definite fluctuating hearing loss. My parents were obsessed with finding out what caused my hearing loss, and I was at the doctor's office a lot. I had a developmental delay as an infant and my parents sought genetic counseling to see what the possibility was for having another delayed baby. This was preschool age (my hearing loss wasn't identified until I entered kindergarten, but it was present before when my parents didn't know what was wrong).

They never got an answer as to the etiology of my hearing loss. The doctors didn't know if I could continue to lose my hearing, so my parents decided our family would learn American Sign Language (ASL). We'd watch videos. I wanted to learn, but I didn't want my parents to think this was "cool" so I refused to engage and just watched silently. My parents let this alone. The doctors, perhaps in some attempt to give my parents something, warned that if I bumped my head, maybe that would further stimulate hearing loss in my good ear, so from grades 3–5 I think I wore a bike helmet in gym class. It was ridiculous, and eventually I refused to carry on this nonsense.

I didn't really get support from a teacher of the deaf—at that time, the hearing levels in the better hearing ear were used to qualify for this support, and I had a normal hearing ear (never mind the profound loss in the other ear), but my mom advocated strongly for support. Once I lost the hearing in one ear, a hearing aid was no longer useful to me, so my option for hearing equipment was headphones (big heavy ones!) coupled with a transmitter I wore around my waist ... my teacher had a microphone. In those days, the systems made a lot of static noise, and I complained, so eventually we went to a sound field system with four speakers in the corners of the room that made the teacher's voice louder for everyone. I was happier because I didn't have to wear any hearing equipment. When I got to high school, it wasn't feasible to have speakers set up in every room, so I was relieved to go without any accommodation— except for sitting at the front of the room. I had an individualized education plan—so before that was officially ended in grade 10, my teachers were aware because I was "flagged" in the system.

My grades were horrible in primary school. The school wanted to hold me back, but my mother would have nothing to do with this. In grades 3 and 4 I was partially mainstreamed. The mainstreaming was a waste of time—the work there seemed misplaced and unconnected to what I did independently in special education. This self-contained class was full of boys—I was one of only two girls ... the other girl was several years older than me. The boys all had bad behavior, and poor academic performance. I was the teacher's pet ... and this was awful! No one else had any hearing loss—I was the only one in my division, the only girl, and one of two students at my elementary school with hearing loss. I don't think this helped. One time, my special education teacher—who was not a teacher of the deaf—was so amazed that I had used cursive writing in my journal one day and had written several pages of narrative. I couldn't understand why this was so impressive—I hadn't put much effort into it. Though (narrative or personal) writing would become a useful tool for me in university, it certainly wasn't anything that I valued in elementary. I found exercises such as journal writing boring. Eventually, I was returned to a normal classroom in grade 5. I struggled academically, but did okay. I was on the honor roll by the end of grade 8. I don't really know how that happened.

I really wanted to attend the deaf program within our school board. I was told I wasn't deaf enough (probably true!), but my longing to be there was more out of isolation than anything else. Things with my peers were awful. The biggest bully in the group of grade 7 and 8 girls was also the one the teachers praised the most. I could never understand this. In particular, I couldn't understand how bullying was allowed to go on, but it did, and it had huge costs for me in terms of self-worth.

At this time, I was away from class one day in 8th grade. I came back to school horrified to find a paper on my chair the teacher had put there, inviting me to a board-wide event for deaf/hard-of-hearing students. Embarrassed, I ripped it up as fast as I could, hoping that no one else had seen it. I didn't want to be like "one of those hard-of-hearing students." Wow, that attitude. Where did it come from?

As I said, I hid my hearing loss in high school from my peers and also from at least some of the teachers. I did use hearing equipment in university, but not consistently at first, and definitely not comfortably. I felt like a victim of hearing loss for a long time—it was hard to accept, until finally, I sought out the opportunity to be with other hard-of-hearing young adults of my own age group. This is what finally made the difference. That first group was diverse geographically and also in terms of hearing loss. A few people signed, and were very comfortable with hearing loss. Others were oral, but still felt so differently (positive) about it than I did.

Growing up, one of my sisters tested in the gifted range, and my parents always said she was much quicker than my youngest sister or myself to understand how a new toy or something worked when we were small. There was a lot of sibling rivalry. My parents always said she was really smart, and didn't have to work for her grades. I had to work for my grades. From these comments, I internalized that I was less smart, but capable if I worked hard.

My mom refused to acknowledge the social limitations or challenges that came with hearing loss. Day after day, she demanded to know whom I played with at recess that day. I went home for lunch most days, and would return to school right before the bell. It was less painful that way. She would express discouragement if I mentioned reading a book, being alone, or with someone in a lower or higher grade by 1–2 years. Eventually, I didn't tell her anything, and spent a lot of time in my room afterschool with the door shut, flung on my bed in tears. Mom said just to keep hanging around with the popular kids in my grade and things would change. They never did. I couldn't hear what they were talking about most of the time. I may have internalized this; I was so introverted and barely said anything in class either. When I did I always got the teacher's attention because speaking was so rare—but this wasn't acceptable to most adults when I was growing up. I don't know if things have changed.

Finally, in grade 13 I sought out my itinerant teacher of the deaf whom had helped with my equipment in elementary (rather than provide direct academic support) and asked to participate in a similar event. I don't remember all the students who attended, but I am still close friends with one of the students in my own grade who attended.

"Knowledge is preparedness"

My parents found out that I had a disability when I was 2 years old. They took me to the children's hospital for a diagnosis—and we were really, really lucky, in that I had a textbook diagnosis. It's important to note that my disability was recognized for what it actually was—a lot of people have been incorrectly diagnosed as having different neurological conditions. My dad and his cousin, who was a doctor, worked through the textbook definition of my disability line by line and figured out what everything meant upfront.

Ever since then, my parents have provided whatever adaptive technologies I needed—but have treated me no different than my nondisabled sister. My parents have encouraged me to explore the world and face it head on, as opposed to hiding me away.

Entering elementary school, I started in kindergarten with a bunch of kids from my neighborhood—one of my neighbors had a chat with the other students about my need to wear sunglasses indoors. Ever since then, all my peers in that class were protective of me—I was never bullied by my peers while I was in elementary school in Canada.

When we moved to Holland, I was temporarily advanced beyond where I should have been for one term—I was in a British school, with students from over 60 nationalities, many of whom came from cultures unaccustomed to interacting with people with disabilities. For the first little while, there were a lot of people asking me—politely and considerately—why I wore sunglasses indoors. I didn't know it at the time, but I was beginning to advocate and communicate about my disability in a way that others would understand. As I got more accustomed to the school, there were a few people who were less than accepting—but this wasn't a cultural thing—and who made fun of me. Although this never bothered me, this was at a time when my family was traveling across Europe constantly, so I wasn't around enough for it to sink in.

This international experience during my elementary education was my compressed "deep end of a cold pool" immersion in multiculturalism and different ways of thinking. I was very aware of the different cultures meshing around me, as opposed to being segregated from each other. As I grew and experienced more things, I don't think I ever accepted that many of these cultures would never accept me as a person with a disability—maybe this is because I don't have a visible disability.

I was one of those kids who wanted to be taken out of my classroom to work with the itinerant teachers—I wanted it when my parents didn't want it, and as I grew older, I wanted that interaction a lot less when it was no longer in my parents' control. As I grew older I realized how much I was missing in the classroom, and how much it impacted the way my peers looked at me, when I was taken out to work with the itinerant teachers. This was never a good thing *for* me—it was a good thing *to* me, and then it stopped becoming even that.

When I attended junior high school, I was back in Canada, and I didn't want to be back. I was attending school with all the people I went to kindergarten with—people who were once my protectors. They had all moved on, and so I never really fit in with them anymore. On our first day of gym class, we were playing volleyball—a white ball flying at your face, coming out of a white ceiling, was something I was terrified of, and so I kept ducking out of the way. All the other players, four on my team and six on the other, who knew me would either exploit or try to compensate for this during the gym class—but there was one student on my team who didn't know me, or why I wore sunglasses indoors, and who finally got fed up and yelled at me, "Maybe if you took off your sunglasses you could see the ball!"

I found this hilarious—and one student on the court next to us promptly replied with, "she can't, she's blind!"

I believe that peers standing up for you when you need them to is much more important than teachers standing up for you in the same situation.

"Disability is not something to be ashamed of"

My parents had a major impact on my trajectory. They always told my brother and I that if we we're going to have a disability it might as well be vision because it didn't prevent us from doing anything. We played hockey, soccer, skied, wrestled, rode bikes, jumped off cliffs, and more. Their mentality was, "you have a disability but it's not something to be ashamed of," it's part of you, and that's the way my brother and I have approached disclosure.

I was exposed to technology from a early age. My dad had computers in the house as early as I can remember. My itinerant teacher really pushed us to learn to type and use a computer from an early age because she knew it would be an important skill for us to have. My itinerant teacher always encouraged me to try new technology to figure out what worked for me. Technology is an equalizer for me. It allows me to read and be a productive member of society.

I had my own computer in the classroom starting in first grade and got my own laptop around grade 6 or 7. Teachers were good at not making me feel out of place because I had these aids, but I still felt awkward using them because it made me different. I still used them because I knew I needed them and without them I couldn't learn. I remember that at the beginning of grade 9 I was very self-conscious that I had a laptop to do my work on. It took me a little bit but I got over it because I needed it.

"What could have been?"

I can't remember the last science class I took. I'm pretty sure it wasn't in high school and definitely wasn't in university. Growing up in Latin America, my educators did not know how to make science accessible to students with disabilities, and so I started behind my classmates in choosing my path in postsecondary education. I didn't have any opportunity to go into the science labs in high school (a totally discriminatory practice)—people were making assumptions about what I could or couldn't do.

I had to learn Braille instead of science, in an "either/or" kind of choice, because there was no other opportunity to learn Braille within the education system. I did not have the equal right or equal opportunity to freely choose what I wanted to do for my career—because of my educators' own choices, whose avenues of possibility were closed to me.

Sometimes I wonder, what would have been my future if I had had the opportunity to learn science?

I don't think the education system—at any level—considers the long-term effects of restricting science education to students with disabilities. What about our participation in society, in those things that we need basic scientific literacy for? What about the effect on our career path and employment opportunities?

Conclusion

While many young people with disabilities share a common experience of overcoming barriers and challenges, sometimes this becomes a narrative that people within disability communities feel they must adopt or fit into. In order to achieve success in STEM, it is important for young people with disabilities to form their own independent self-narratives and identities, driven by their experiences and aspirations.

Educators/instructors fall into three broad categories. This transcends across educational level and type of instructor. These types are:

1. The instructor who is interested, engaged, and sees instructing a student with a disability as a unique learning opportunity that will enable the faculty member to become a better instructor. This type of instructor is very enthusiastic about the idea of being creative and flexible.
2. Those who are outright negative, hostile, and make it difficult to receive support/put up road blocks. This may be in the form of not providing materials in an alternative format, or just simply an unwillingness to think creatively and critically about how to adapt course materials.
3. Those who ignore, pretend we, as students with disabilities, don't exist and pay no attention at all. This last category is actually the most discouraging and damaging, because there is no effort to even acknowledge the student is present.

A recurring theme, which is perpetuated by some disability-related services staff and educators, is a de-emphasizing of certain subjects that are perceived by some to be "less than necessary." Some people with disabilities even have educational experiences that have been so whittled down from a more conventional education that they are essentially left only with subjects that underprepare them for postsecondary education in math or science. An example might be an education where grade-appropriate math or science classes have been replaced with mobility training, independent living skills, Braille, etc. While these subjects are certainly important, this discussion raises an important question. Aren't math and science important as well? Even if a person chooses not to pursue a STEM field, won't mathematical skills be useful? Calculating student loans, mortgage payments, interest, budgeting, etc., all require basic numeracy, and the ability to understand numbers and mathematical models. Attitudes from outside as well as within the disabled community range widely.

The problem associated with drastically altering a curriculum is the fact that a student lacking in an essential field of study has fewer options after secondary school. Some educators hold the view that the ultimate goal is to "get a student through" their secondary education criteria. Others hold the view that a student's exposure to all STEM fields should not be dictated by the presence of a disability. The latter viewpoint produces students who have significantly more options at the postsecondary level.

What factors have shaped these two mindsets? Differing perceptions of disability in education are created by different priorities and objectives by educators. One school of philosophy teaches that the ultimate goal is to teach a person the basic living skills necessary for living as a disabled person. The other teaches that adaptation and ingenuity are to be implemented to existing environments to allow for accessibility to disabled persons even in areas of study and occupations that are not traditionally occupied by those with disabilities.

Regardless of the viewpoint of the educator, the student in question is almost always underconsulted in regards to making decisions regarding curriculum. At the postsecondary level, students with disabilities have just emerged from an environment where they may have been instructed by parents, educators, or disability service providers along every step of their educational path. Whether the student prefers this type of guidance or not is not typically a matter open for deliberation. When such students reach the college or university level, they are suddenly responsible for navigating uncharted waters and are expected to advocate for their own needs. It is most typically at this point where students who have been adequately prepared to adapt in response to new challenges become apparent.

The role of accommodations also comes into play in discussions between members of each camp of thought. Those who would push for an emphasis on an education that revolves around the acquisition and mastery of independence and living skills might argue that excessive accommodations would make a gainful career in most STEM fields too difficult to orchestrate. While it is true that trailblazing is no easy feat, the notion that accommodations cannot be efficiently integrated into a work environment is shortsighted. Acknowledging the fact that it is not the aspiration of every disabled person to end up in a STEM career, those who are willing to put in the time and work should not be dissuaded on the basis of needing to facilitate technologies and other accommodations that allow for independence and efficiency. After all, what is seen as "excessive" to one generation may be seen as commonplace in the next. This reality underpins the need for widespread implementation of accessible laboratories.

Societal attitudes regarding disability can and do impact the lives of people with disabilities on many levels. Perceived limitations, reinforcing stereotypes, a lack of communication with disability advocacy groups, and more all place additional barriers along the pathway of a student with a disability in STEM fields. It has been shown that societal expectations and stereotypes have impacted the gender ratio in STEM fields. It is clear that reinforced stereotypes have the power to steer willing students (either consciously or subconsciously) away from STEM fields of study, and consequently away from STEM careers. Such societal pressures can include an aversion to adaptive technology, incorrect assumptions regarding the mental capabilities of students with a disability/disabilities, incorrect assumptions regarding the value of a student with a disability/disabilities experiencing adversity related to his or her field of study, and the perception that students with disabilities may be emotionally underprepared for careers that require rigor and persistence.

Ultimately, what does the inclusion of people with disabilities in STEM fields mean? It means opportunity for both the employees as well as the employers. By setting a new normal for the presence of qualified disabled persons in nontraditional fields, the benefits extend not only to disabled people, but also to their respective fields by contributing new influxes of innovation and talent. By increasing the expectations for qualified individuals and challenging dehumanizing perceptions, an increase in the presence of lateral thinking will take place in fields that are no longer bound by conventional modes of analysis and problem-solving.

Key role of education providers in communication with students and service providers

Chapter Outline

Introduction   59
First contact: how faculty find out students with disabilities are in their courses   60
Faculty responsibilities around communication and content delivery   60
Scope of the teaching team   62
Some typical situations   62
  Case study #1: undergraduate lab course-based (with pitfalls)   62
  Case study #2: undergraduate lab course-based (ideal scenario)   63
  Case study #3: graduate research environment (with pitfalls)   64
  Case study #4: graduate research environment (ideal scenario)   65
Conclusion   66

"You're the first blind geneticist I ever heard of."
    Almost twenty years after first hearing those words – in the third year of my undergraduate studies, in the first lecture of my introduction to genetics laboratory, spoken to me by that course's faculty coordinator – they continue to resonate. In many ways, for better or for worse, they defined what my career in science and higher education would become. They also illustrated the mountain of faculty attitudes to be scaled, and the potential difficulties in faculty engagement during my time as a university student.
    To her credit, this faculty member – once she got over her surprise at meeting someone who wanted to do what common wisdom suggested was impossible at the time – was very helpful, and went out of her way to ensure my accommodation needs in that course were appropriately addressed. That meant a great deal to me – I was dealing with a lot of "culture shock" because it was a new environment, a new program, a new set of students to get to know (none of which were disability-related issues), compounded by having to negotiate my accommodations in a new setting, and having to do so in what I eventually realized was a hostile environment: I was in a program where the vast majority of faculty did not believe I belonged. At best, they were indifferent to my needs. At worst, they were outright oppositional.

Creating a Culture of Accessibility in the Sciences. DOI: http://dx.doi.org/10.1016/B978-0-12-804037-9.00005-X

And yet, even there, I met faculty members like this course coordinator who were willing to engage in constructive dialogue. Or like the faculty member who expressed commiseration about the accessibility of the campus and the lab environment because of her temporary disability the year before. Or like the program coordinator who admitted me to the genetics program in the first place, who saw potential in my academic record and mindset, and was willing to nurture that.

Those were the faculty members whose examples provided the courage to continue, particularly when others were much less than communicative or accommodating. This was a trend I saw over and over again throughout three postsecondary degrees and two postdoctoral fellowships across eighteen years of training – and suggests, sadly, that we still have a significant way to go to ensure inclusivity of the STEM disciplines for persons with disabilities.

The best faculty I worked with – the ones I learned the most from as role models and mentors – modeled how to engage in constructive and effective dialogue, not just with me, but with the disability services office as well. Indeed, many of the suggestions and highlights contained in this chapter were drawn from my experiences with these individuals. The worst faculty I worked with proved themselves actively uninterested in any constructive dialogue – I learned a great deal from them also, about how not to engage with students and service providers. Those lessons made their way into this chapter as well.

I found, though, that most faculty were willing to be of assistance, but often did not know how. In a setting where the student is feeling their own way through learning their craft and in many ways "swimming up-river" against common perceptions about science and disability; and, where the service provider is as well intentioned but unfamiliar with the technical aspects of the field the student is in, the lack of knowledge on everyone's part is itself a significant barrier to communication and collaboration. When everyone is willing to help, but no one knows exactly how, everyone feels lost, inadequate, stuck. In my case, faced with these situations, I learned to muster up my courage and took charge of the dialogue, ensuring it happened and directing it to a positive outcome.

Not every student would choose that route. In those circumstances, perhaps it should fall to the faculty member to step forward and foster the dialogue, to become an advocate for the student and their success. The importance of faculty to student success is particularly understated and underappreciated – but is a role that is inextricably tied to the faculty member's role as educator and facilitator of learning.

Introduction

Faculty members play a particularly important role in creating a welcoming and inclusive teaching environment, as "University faculty are on the front line of ensuring that students with disabilities receive a quality postsecondary education" (Zhang et al., 2010, p. 285). Faculty "willingness to provide reasonable accommodations and related supports to students with disabilities is essential to the success of these students" (Zhang et al., 2010, p. 285). Furthermore, the quality of the services that faculty provide plays a significant role in the ability of these students to complete their higher education (Zhang et al., 2010, p. 285).

Cook, Rumrill, and Tankersley (2009) also note that "the success of any [university] student, particularly in the academic realm, is to some degree determined by the type and quality of interactions" with faculty (p. 84). Faculty members make crucial contributions to the campus climate and the learning environment and thus their priorities and behaviors "are important determinants of the quality of higher education for students with disabilities" (Cook et al., 2009, p. 84). Johnston and Doyle (2011) found that students with disabilities felt included, and better able to succeed, when faculty demonstrated an understanding of disability and accommodation, and then implemented inclusive teaching practices. Also identified to be of benefit was that faculty would allow a student with a disability to try a task before making a determination around his or her inclusion or participation.

When faculty actively engage in using "effective educational practices, students will engage in them and benefit in desired ways" (Kuh, Laird, & Umbach, 2004, p. 30). In particular, faculty need to be flexible, solution-oriented, and creative in designing tasks, tools, and the laboratory environment, in order to allow students with disabilities to actively participate in science laboratories (Heidari, 1996; Langley-Turnbaugh, Murphy, & Levine, 2004). As Miner, Nieman, Swanson, and Woods (2001) highlighted, "teaching chemistry to students with disabilities can provide new opportunities for teachers and students to use their creativity in the classroom and the laboratory" (p. 95). For example, Supalo, Mallouk, Rankel, Amorosi, and Graybill (2008) describe many low-cost laboratory adaptations available for students with low vision or who are blind, including tactile adaptive technologies, technological solutions, 3D models with drinking straws, and tactile 2D models.

Moon, Todd, Morton, and Ivey (2012) note that while these accounts of accommodation solutions "may hold promise, the success of such solutions relies heavily on the personal motivation of individual instructors to meet the accessibility needs of their students" (p. 30). Furthermore, there is as yet no consensus on "a standard design for an accessible lab in any science, technology, engineering, and mathematics (STEM) field or at any educational level [and there are] few examples of accessible versions of advanced laboratory equipment … [Therefore], neither a fully accessible lab nor other off-the-shelf accommodations may be available [to the student]. While this gap poses challenges for STEM instructors, it also represents an opportunity for faculty to develop their own solutions and adapt them to their particular needs" (pp. 30–31).

First contact: how faculty find out students with disabilities are in their courses

There are a variety of ways in which professors find out that they will be teaching a student with a disability. Some professors receive a note from their institution's coordinator of services for students with disabilities (the title of these individuals and offices differs from place to place—their function is to assist students in obtaining the special support or services they may require). A short pamphlet about the nature of the student's disability and about recommendations to professors may accompany the note. Sometimes the student may contact the professor before registration or before the start of classes. However, in most cases the student simply appears on the first day of class and the professor becomes aware of the student's presence if the disability is visible.

With advance notice, professors might wonder about how to accommodate the student's needs, what types of changes to course structure may be necessary, and how the student will cope. If the student contacts the professor before registration, what should he or she say if it is felt that the student's disability will make it impossible or extremely difficult to succeed in the course? What accommodations are reasonable to make and what kinds of adjustments are unwarranted? If the student simply shows up on the first day of class, what should you do? Do you approach the student to ask whether special consideration is needed or should you leave it up to the student to initiate dialogue?

Faculty responsibilities around communication and content delivery

While the faculty member is knowledgeable about the essential content that must be delivered and the essential skills that must be learned or maintained, the disability services office has the information about the impact of a disability on learning. Thus the interaction between faculty and the disability services office "cannot be partitioned; collaboration and communication between the two groups is vital" (Madaus, 2000, p. 19). It is also important for faculty and students with disabilities to develop a good working relationship with each other, and to keep communicating as the semester progresses; this is especially important in laboratories where different tasks arise throughout the course. The office for students with disabilities can help faculty with locating lab and classroom supports, organizing the transcription of materials, and planning for accommodations. However, the faculty member will often see the student more frequently than would the disability services office, and the student will have the best knowledge about his or her own disability (Miner et al., 2001). Effective communications among all parties is thus critical.

Students themselves are often the best resources of information about their needs and the accommodations that will best help them succeed in the classroom or lab. Therefore a faculty member's concerns surrounding what is working well and what is not in the lab can first be fielded to the student for feedback and suggestions (Miner et al., 2001). Certainly, the office for students with disabilities needs to be involved in the accommodation process, but as the faculty member and student have a more frequent interaction, concerns can be resolved more quickly if they are addressed as they arise in the lab: "It is very helpful to be *shown* [emphasis added] how to use the equipment needed for labs rather than just having to rely on written instructions in the manual or on the equipment. Also, it is helpful if instructors/Teaching assistants (TAs) are available during equipment use to answer questions/help solve problems" (Johnston & Doyle, 2010). Additional helpful tactics include having time outside of class for the student and instructor to meet and revisit techniques that were covered. Often, questions are asked by the student to seek clarification in order to avoid becoming lost during the lecture or demonstration.

In addition to providing course accommodations, it is important for faculty to be aware of the accommodations that will permit students to participate fully in the lab. A lab assistant can be employed to carry out only those tasks that a student with a disability is physically unable to accomplish (Neely, 2007; Pence, Workman, and Riecke, 2003). For example, a lab assistant may read out measurements to a student who is blind, or measure and weigh materials (as instructed) for a student with dexterity or mobility challenges. A faculty member can assist in locating a lab assistant by suggesting a student who previously completed the course and has mastered the required experimental techniques. At the graduate level, lab assistants can also be skilled technicians hired for this specific purpose. Students with disabilities in lab courses are learning alongside their classmates and therefore cannot always rely on them to fill the role of lab assistant effectively (after all, the classmate may be prone to making errors the student with a disability may not have made themselves). However, engaging a classmate to perform this task should not be ruled out, and in fact, according to disability services providers, it is the most frequent type of accommodation arrangement.

Faculty members should identify potential safety hazards in the lab and suggest options for how to handle an emergency, so the student and the laboratory assistant can establish general procedures with each other and determine if, or under what circumstances, the student with a disability may require assistance (e.g., if a student cannot hear a safety alarm). It is also important to confirm a student's understanding of the safety instructions, which may be especially important for some students such as those on the Autism spectrum (Hughes, Milne, McCall, & Pepper, 2010). The student should also be given an opportunity to express any concerns about his or her safety in the lab (Pence et al., 2003).

Safety and liability concerns should be discussed with the student and disability services provider at the beginning of the term, so that the former has time to become familiar with lab safety procedures. The instructor should prepare for possibilities like the evacuation of students with mobility and sensory disabilities from the building during emergencies. This is especially important to consider in advance of a course.

It is important to remember that restricted laboratory use does not preclude a productive scientific career. With the right instruction, mentoring, and accommodations, a student with a disability can be just as successful in the laboratory as any other student. In sum, while accommodating students with disabilities is a legal obligation, what is most important is the spirit of the law: providing appropriate accommodation and ensuring an accessible environment makes for welcoming and inclusive education without compromising the essential course curriculum.

Scope of the teaching team

At the undergraduate and graduate levels of postsecondary education, science is often "team-taught." A course coordinator may arrange for a series of course instructors, each a subject matter expert in his or her own area, to teach various sections of a single course. For each content section of the course, the essential requirements (see chapter 10: Essential requirements and academic accommodations in the sciences) may also be subtly or overtly different. Courses with lab or tutorial sections (or both) will often have TAs (sometimes different TAs for different parts of the course), who may teach in the lab or in the tutorials. Again, there may be substantively different essential requirements in different parts of the course—or different essential requirements in the lab compared to the lectures. For example, the defined essential requirements for an introductory physics lab course covering force and motion are likely to be very different from those for the same course when the students are studying sound waves.

In the context of a teaching team, it may not be intuitively obvious to the student who to contact with respect to his or her accommodation needs; indeed, the first point of contact may not be the best. Additionally, the person with the greatest level of inter-action with the student (e.g., a TA) may not have the authority to provide assistance. This speaks to both the need to train all parts of the teaching team around accessibility and accommodation, as well as the student's openness to multiparty dialogue around these issues.

Some typical situations

Successful engagement and dialogue among faculty, students with disabilities, and the disability services provider is dependent on all three parties being open to and willing to engage in collaboration and correspondence. All three parties bring something to the table—the faculty member, the detailed technical knowledge of the discipline and program requirements; the services provider, the knowledge of accommodation processes and practice; and, the student, an understanding of his or her disability and accommodation needs. The scenarios below illustrate some common best (and worst) practices in the engagement among students, faculty, and service providers.

Case study #1: undergraduate lab course-based (with pitfalls)

This case scenario outlines the experiences of an undergraduate student as he or she attempts to set up and navigate the accommodations for the undergraduate teaching lab environment:

- The student does not communicate with his or her disability counselor until after the start of the lab course.
- Their first meeting occurs after 1–2 weeks in the term, and is rushed because the student has not planned ahead for the conversation.
- The disability counselor suggests accommodations, and the student agrees to an accommodation set that may not be the most appropriate due to a lack of advance preparation and research into the course and its objectives.
- The disability counselor communicates about the student and his or her accommodation needs with the faculty member running the course after the first meeting with the student; the faculty member reviews the communication cursorily and chooses to not intervene without contact with either the student or the disability counselor.
- The student, trusting the process that the disability office sets out, chooses to not contact the faculty member to discuss his or her accommodation and accessibility needs or potential solutions.
- Throughout the course, the student has little communication with the faculty member or the disability counselor, and chooses to carry out his or her course work with the supports initially put in place.
- The course coordinator notes the student's poor performance, but chooses to not engage with the student to understand his or her concerns and challenges, attributing poor performance to a lack of natural aptitude for the course material, or a lack of engagement on the student's part.
- There is no ongoing evaluation of the accommodations and their effectiveness, and no ability to capture and respond to concerns or emerging issues.

Outcome: The student receives improper accommodations for a science lab course, and has no interaction with his or her course coordinator or disability counselor throughout the course. As a result, the student is frustrated that accommodations are not working, may do poorly, and may reconsider his or her engagement with lab courses, the program or discipline, or even the sciences generally in the future.

Case study #2: undergraduate lab course-based (ideal scenario)

This case scenario outlines an undergraduate student's journey as he or she works with a faculty member or educator and disability support staff to establish appropriate accommodations in an undergraduate teaching lab environment. In an ideal situation:

- The student meets prior to the start of the term (2–4 weeks) with his or her disability support staff/advisors and establishes a framework around needed accommodations, support and strategies for learning based on his or her specific disability, and the expectations the student has for the course.
- The counselor and the student work through this framework to define an accommodations plan, which is then formalized and communicated by the counselor to the faculty member or educator in charge of the lab course. This happens prior to the start date of the course.
- The student arranges to meet with the faculty member serving as course coordinator to discuss his or her accommodation needs in the context of the course and the prescribed accommodations plan that is put in place by the disability services office.
- The faculty member and student propose modifications and creative solutions to outstanding accessibility concerns based on their discussions, the faculty member's knowledge and

technical expertise, and the student's own experience, and communicate these back to the disability counselor.

- The disability counselor, working with the faculty and student, modifies the accommodation plan appropriately, and all three work together to ensure its proper implementation.
- For team-taught laboratory courses, the student, disability counselor, and faculty member agree on the appropriate level of information to be shared with the remainder of the teaching team (including TAs), as well as on who is responsible for sharing this information and working most closely with this group.
- In the event of challenging situations that were not initially foreseen, the student communicates first with his or her course coordinator prior to speaking with the disability counselor.
- Throughout the length of the course, the faculty member, student, and disability counselor meet to review progress and success of the accessibility solutions and accommodations, as well as to address any nascent barriers that may have come up.

Outcome: The student does well and develops a really strong relationship with the faculty member, gaining a mentor in the process (see chapter 13: Faculty mentorship of students with disabilities in the sciences). The faculty member, having engaged with the student and disability counselor, knows what works best for that student for future reference, has a deeper understanding of disability and accessibility considerations in his or her course, may take the opportunity to make the course material more inclusive in future years, and can be more responsive to future students with disabilities and their accessibility considerations. The student, having gained confidence in the technical aspects of the course material, as well as in his or her own self-advocacy skills, continues in the program.

Case study #3: graduate research environment (with pitfalls)

This case scenario outlines the experiences of a graduate student as he or she attempt to set up and navigate accommodations for entering a research-stream graduate program in the sciences:

- The student receives acceptance to his or her graduate program of choice and an invitation to join the laboratory research group.
- The student does not communicate with his or her disability counselor until the after the start of graduate school.
- Their first meeting occurs after 2–4 weeks into the graduate program, and is rushed because the student has not planned ahead for the conversation.
- The conversation between the student and disability counselor focuses exclusively on the course work for the student, and not on his or her research environment, academic employment, and other accommodation needs. The student is unaware and does not believe that these parts of the graduate program are under the purview of his or her disability counselor.
- The disability counselor suggests accommodations, and the student agrees to an accommodation set that may not be the most appropriate due to a lack of advance preparation and research into the graduate program and its objectives.
- The disability counselor communicates about the student and his or her accommodation needs with the faculty members running the student's course after the first meeting with the student; the faculty members review the communication cursorily and choose to not intervene without contact with either the student or the disability counselor.

- The student, trusting the process that the disability office sets out, chooses to not contact the faculty members in their department to discuss his or her accommodation and accessibility needs or potential solutions.
- Throughout the course work the student has to undertake, the student has little communication with the faculty members or the disability counselor, and chooses to carry out his or her course work with the supports initially put in place.
- There is no ongoing evaluation of the accommodations and their effectiveness, and no ability to capture and respond to concerns or emerging issues.
- The student chooses to not engage with his or her supervisor or thesis committee, and does not inform them of accommodation needs or accessibility considerations.
- The student chooses to attempt self-accommodation in the context of his or her graduate research and does not reach out to solicit advice or perspective from the supervisor.
- The student does not explore with his or her supervisor or thesis committee what they could potentially offer in the way of funding, resources, advice, or accommodation.
- Milestones throughout the graduate program are attempted by the student without appropriate discussion or accommodation.

Outcome: The student's course work at the graduate level starts out at a high level, but ultimately may plateau or suffer as frustration builds around the other aspects of the graduate research environment. No supports to aid the student are put in place in the research lab or other research settings, or in the academic employment context, and no dialogue with the supervisor or thesis committee around these crucial issues occurs. The student's performance in the research environment plateaus and begins to suffer faster than his or her course work. The student is unable or unwilling to communicate with his or her supervisor or thesis committee around disability and accessibility issues, while trying to maintain a functional research relationship.

As the student's performance continues to degrade, trust issues begin to set in. Although the student may ultimately pass his or her examinations and earn a degree, the relationships with the department, thesis committee, and supervisor are not strong. It is also likely that the student's performance issues would be called out by the supervisor and thesis committee, leading to potential additional mental health considerations (in the best-case scenario) or outright termination of the student from the program and laboratory (worst-case scenario).

Case study #4: graduate research environment (ideal scenario)

This case scenario outlines a graduate student's journey as he or she works with the thesis supervisor and/or thesis committee as well as disability support staff to establish appropriate accommodations in the context of the graduate research laboratory environment in STEM. In an ideal situation:

- The student receives acceptance to his or her graduate program of choice and an invitation to join the laboratory research group.
- The student meets prior to the start of his or her program (2–4 months) with disability support staff/advisors and establishes a framework around needed accommodations, support and strategies for learning based on his or her specific disabilities, and the expectations the student has for the graduate program.

- The student discusses the application of accommodations to the course work, research, and academic employment environments with his or her disability counselor and gains an understanding of the differences in accessibility considerations in each of these arenas.
- The counselor and the student work through this framework to define an accommodations plan, which is then formalized and communicated by the counselor to the student's supervisor. This happens prior to the start date of the program.
- The student arranges to meet with his or her supervisor (and potentially the graduate coordinator as well) to discuss accommodation needs in the context of the graduate program prior to communication of the formalized accommodation plan from the disability counselor.
- The student and supervisor discuss relevant issues around funding, program, and degree requirements and integration into the lab or research group setting. The student and supervisor identify appropriate sources of funding (e.g., grants and awards) that could be applied for to support the student in the lab or research group. A plan is put in place to apply for this funding over the duration of the program.
- The supervisor and student propose modifications and creative solutions to outstanding accessibility concerns based on their discussions, the faculty member's knowledge and technical expertise, and the student's own experience, and communicate these back to the disability counselor.
- The supervisor becomes a champion of the student's accomplishments and accommodations to his or her faculty colleagues and the university as a whole.
- The disability counselor, working with the supervisor and student, modifies the accommodation plan appropriately, and all three work together to ensure its proper implementation over the duration of the program.
- The student, disability counselor, and supervisor agree on the appropriate level of information to be shared with others including the thesis advisory committee, department, collaborators, and those responsible for the academic employment of the student (if necessary). Additionally, the method of disclosure (i.e., through which party will disclosure be allowed) must be discussed and agreed on in advance.
- In the event of challenging situations that were not initially foreseen, the student communicates first with his or her supervisor prior to speaking with the disability counselor.
- Throughout the length of the program, the supervisor, student, and disability counselor meet to review the progress and success of the accessibility solutions and accommodations, as well as to address any nascent barriers that may have come up.

Outcome: The student has a healthy and strong relationship with his or her supervisor. While this relationship may be tested by research-related issues, disability is not a factor in the stress in the relationship. The student graduates having successfully completed his or her program of study, and has lasting mentorship and collaboration relationships with the supervisor, thesis committee, peers, and members of the department.

Conclusion

While it is ultimately up to the student how much he or she wishes to engage with faculty around accessibility and accommodation issues, there are steps that faculty members can undertake to ensure that students know that they are open and receptive to such dialogue.

- You (the faculty member) should approach and talk with students with disabilities (after they have identified themselves to you) early in the semester or relationship about working together on accessibility and accommodation issues—during the first few days of classes and shortly after first meeting.
- If students don't approach you, don't wait—approach them yourself and let them know that you are open to communication.
- Do not single out students for special attention in class or the research setting, because this could embarrass them. Speak to them privately about problems or issues related to the disability.
- Discuss course or research issues related to the disability and identify potential problems. Talk about what the student can and cannot do in your course and discuss what adjustments or modifications you might make in your teaching style and in your evaluation procedures.
- Encourage students to keep in touch with you during the term so that problems can be solved as they arise and so that crises can be averted. Let students know that you are available to meet with them.
- In matters where the disability is not an issue, treat the student as you would any other student.
- Understand that the student is encountering new situations during his or her undergraduate and graduate studies and that he or she may not always have the answer or solution to issues regarding accommodation needs. It is important to let the student know that you understand that new situations require trial and error and that accommodations can be refined throughout the duration of the class, semester, course of study, etc.
- Keep in mind that, like students without disabilities, students with disabilities may struggle with certain concepts and/or skills and techniques for reasons unrelated to their disability/ disabilities. Nondisability-related obstacles can and will be encountered just as they are with nondisabled students and therefore not every difficulty should be "blamed" on a person's disability.

Part III

Student as Educator

Disclosure in the sciences

Chapter Outline

Introduction 73
Definition of disclosure 73
Disclosure of accommodation need vs. disclosure of disability 74
Process of disclosure 75
The choice to disclose 75
Self-advocacy and disclosure 76
What students should know prior to disclosure 77
Timing of disclosure 77
A rubric for disclosure 78
Identifying the right players 78
Impact of disability types on disclosure 79
Pros and cons of disclosure 79
Implications of disclosure in laboratory and practical space environments 80
Other factors affecting disclosure 80
Conclusion 81

"I have a visual impairment and I'm in your class."

For a long time, I found those easy words to utter. Throughout my undergraduate degree, I was only too pleased to try to start that conversation with my instructors and faculty members. In retrospect, I wasn't prepared for the range of responses that utterance would evoke, from helpfulness, to indifference, to outright hostility. As an entry point to a conversation, honestly, that method of disclosure didn't always work. While some would be willing to engage in a conversation around my accommodation needs, others were more perfunctory or dismissive. "I have this letter from disability services. I'll read it later." Or, "I did what the request from the service provider said, I have no more responsibility to you."

Not everyone considers their disability akin to a personal characteristic, like hair color or skin color or height or age. I do, however; for me, being visually impaired was the same as being four or five years younger than my cohort throughout my career. It just was, so figuring out how to shape the conversation differently when I began to have it with faculty was something of an exercise. Of course, if everyone had the same perspective I did, then there wouldn't have been a need for the conversation in the first place.

Theoretically.

Creating a Culture of Accessibility in the Sciences. DOI: http://dx.doi.org/10.1016/B978-0-12-804037-9.00006-1

The problem was that not everyone had the same perspective I had – never mind faculty members, not every person with a disability has my perspective. As I got older, and as I encountered negative reactions to my "disclosure conversation," I began to understand why.

So, I tried to change the conversation.

"I need some accommodations to function well in your class."

Truth be told – and this isn't a pleasant truth – that doesn't work a whole lot better. We're told that it's supposed to be a better entry point to a conversation with faculty, that no one can mandate us to disclose a disability, but that we ought to consider talking about our accommodation needs. That's a much harder conversation to start – we're less familiar with it, after all. And the most common response I received was:

"OK. Why?"

"Well, I have a visual impairment..."

Wait a minute. Isn't this where we started?

The thing is, I am extremely comfortable with my disability and accommodation needs – and have been for a very long time. In a community setting, I'd found that I had no problem disclosing and (if necessary) receiving necessary accommodations. And, of course, everything in high school was negotiated on my behalf. So what was the real problem in the postsecondary setting?

The challenge, as I observed it, was that faculty just weren't expecting someone with a disability in their science classes. In my undergrad, the likelihood of meeting another person with a disability in any of my science classes – lectures, tutorials, labs, didn't matter – was vanishingly small.

In fact, it never happened.

So, I was terrifically unusual. Outside everyone's experience, in fact. That's when I ran into the bias of others' total absence of experience: If it hadn't happened yet, it must be because it wasn't possible.

How to navigate the disclosure conversation with that mental framework in mind, then?

One, I recognized I required some level of accommodation or technical assistance to achieve my goal of being a scientist. Trying to do that on my own was not going to work. (Not for me – the accommodation set I needed was just not something I could have negotiated on my own.)

Two, I recognized that disclosure – either of disability or of accommodation need – wasn't enough. I had to be prepared to come with solutions, or at least options to consider in conversation. After all, I knew myself best, and being able to offer alternatives was the easiest way around the knee-jerk "No, I can't do that."

Three, I needed to drive that conversation. This wasn't about persistence as much as it was about ownership. At some level, it was about taking personal responsibility for what I needed, and recognizing that, legal obligations or no, no one else was going to help me get it, or achieve my goals for me.

Four, "No" was not a viable answer, neither was failure an option.

Five, strategy was key. What to say, when to say it, how to say it, who to say it to – ll this required planning and thought, and wasn't something to merely rush into.

And, six, success speaks louder than anything else. I wanted accommodations? I would demonstrate that I was good at what I wanted to do, and that the accommodations, while necessary, weren't a crutch. I also would demonstrate that my own innate skill set came through, irrespective of what accommodations I needed. In that way, my accommodations served a purpose of enabling my latent aptitude and skills, which is fundamentally different than having my accommodations create my skill set for me. This latter perspective was the fear that I began to understand others had.

Disclosure is such a personal thing. Many people will shy away from it for legitimate reasons. Navigated properly, and done well, disclosure is a powerful and necessary tool in our ability to advocate for ourselves and our goals and aspirations to become successful scientists.

Introduction

Disclosure is a process or convesation that is often misunderstood by students, faculty, and departmental staff. When must a student disclose his or her disability? To whom? To what level of detail? What kind of information? It is a process that varies for each student, depending on his or her disability, program, and overall comfort level, among other factors. This chapter will walk through what it means to disclose, the choice students have, the process they must undertake, and the associated aspects students must consider.

Definition of disclosure

Disclosure is "an approach used to inform an [employer, prospective employer, or postsecondary institution] of a disability that needs to be addressed and accommodated" (Ryerson University, 2016).

Disclosure in the academic environment means making it known to various parties that the student has some form of a disability and therefore requires accommodations in order to succeed to his or her full potential. Disclosure can take many forms and is highly dependent on an individual's specific disability.

Disclosure is a highly personalized and sensitive topic that is often misunderstood. There is no right or wrong way to go about the disclosure process. It varies for everyone, based on disability, the program of study you are planning on taking, and the type(s) of accommodation being sought. Some students may choose not to disclose to their faculty or even their accommodation specialists, which is perfectly acceptable. However, it is important for the student to understand the pros and cons of the choice to disclose (or not). Information must be made available to allow the various parties to understand the ins and outs of disclosure, which can be beneficial in opening up a conversation surrounding it. Ultimately students must do what is right for them and their education. This chapter aims to make the process clearer and provide the information needed to help people to make informed decisions.

There are many situations in which disclosure may take place, and thus it may be handled slightly different in each situation. For example, a student may disclose the need for a notetaker in a classroom setting, or for an assistant in a lab, fieldwork, or placement setting. Disclosure is also inherently different at the graduate level, due to the individualized aspect of many programs. In each situation the student must decide whether or not they wish to disclose, and if they choose to disclose, figure out to whom and what to disclose, which will be explored as this chapter progresses.

Disclosure of accommodation need vs. disclosure of disability

Something that is often misunderstood by students and educators is that there is a difference between disclosing a diagnosis and the need for an accommodation. With an increased number of students with diverse needs, understanding disclosure is critical. In order to receive disability-related accommodations the student is only required to disclose his or her actual diagnosis to the disability services office (DSO). The diagnosis is confidential and is not shared with faculty, but the accommodation needs can be shared (Johnson, 2006). The student must also provide proper documentation from a qualified professional. This is also true for receiving disability-related funding. However, students are not required to disclose their actual diagnosis to their professors, teaching assistants (TAs), or department. Students must only disclose the need for an accommodation to those parties, providing they have disclosed this need to the DSO.

An accommodation need could be anything from a notetaker to a lab or field assistant. This need is often handled through the DSO and only requires the professor's approval. In some cases it requires the professor or department to work collaboratively with the student and DSO to ensure the best plan is put into place. The decision to disclose, even to the DSO, is a highly personal one. Some students, especially those with invisible disabilities, make the decision not to disclose out of fear. For those with visible disabilities that decision can be taken out of their hands, but the requirement remains the same in terms of simply disclosing an accommodation need. The next section explores the process a student must undertake if they choose to disclose.

Process of disclosure

In order for students to receive formalized accommodations, they must first disclose to their DSO. This includes providing the proper documentation completed by an appropriate professional, often followed by a meeting with an advisor to discuss the needed accommodations. Students may need to disclose their accommodation needs to several individuals (faculty, department heads, etc.) before the accommodations can be discussed by a team. The student and DSO must work within the policies and procedures (if extant) of academic departments, faculties, and the institution in order to organize accommodations.

Students may then approach each individual professor to explain their need for an accommodation. Students may also be responsible for explaining their needs to TAs, departments, and others they must interact with. However, aside from disclosing his or her diagnosis to the DSO, it is up to the student whether to disclose, and if so how much to disclose. Students are also only required to disclose to the DSO if they want to go through them for accommodations. They may choose to self-accommodate, if they do not feel comfortable disclosing, or feel that they may be in a position to manage their accommodation needs on their own. It is important for students to be able to navigate the system while ensuring they feel comfortable with the process and what they are disclosing.

The process of disclosure can vary depending on the situation and the accommodation being sought. For example, students in more practical settings may be more likely to also have to disclose to TAs, whereas those in arts-based programs may be more likely to have to deal with individual professors. Some students may not feel comfortable with the process of disclosure, due to either not understanding it or feeling it violates their privacy. The next section explores the choice the student has in terms of disclosing.

The choice to disclose

There are many factors that can influence a student's decision to disclose. This includes type of disability, program, individual factors (e.g., how comfortable the student is talking about his or her disability), etc. Students must weigh their lists of pros and cons of disclosure before making a decision. Students can choose not to disclose, but this could cause issues if a student has a crisis situation and thus losing the choice about whether to disclose or not. A useful tool for weighing the pros and cons of disclosure is a thought frame around identifying the benefits and drawbacks to disclosure (see the disclosure "matrix" proposed by Roberts, 2014).

One of the biggest factors influencing the decision about whether to disclose is the type of disability. Unique challenges are posed for both those with invisible and visible disabilities. A student may fear stereotypes and stigmatization, that he or she may further be seen as "other," separate from peers. The choice to disclose is also heavily dependent on the student's comfort level and knowledge surrounding his or her disability and accommodation needs. In order to adequately explain their needs, students must be given the tools they need to disclose. Students right out of high school often do not have the tools they need to get their needs met, nor are they fully aware of the resources available to them. Students must also know the various parties they can disclose to, including the DSO, professors, TAs, and others involved in their education. Regardless of when and to whom a student may choose, or not, to disclose to, it is a highly personal decision.

The choice about whether to disclose is also heavily dependent on the program in which a student is enrolled. Students enrolled in science- or professional-based programs may be less inclined to disclose due to the fact they fear added stigma within their program. They may also fear being seen as less competent than their peers to complete the program. This is also true for students enrolled in graduate programs. Students are often unaware that, even at this level, the DSO is still able to provide accommodations or do not seek them out due to fear of possible ramifications because of the competitive nature of some programs.

Self-advocacy and disclosure

Self-advocacy is a major issue for many students, and is the reason many students choose to disclose (Johnson, 2006; see chapter 3: Barriers faced by students with disabilities in science laboratory and practical space settings). Self-advocacy and appropriate disclosure is almost always the responsibility of the postsecondary student. The student knows his or her disability and needs best, and is in the best position to speak about accommodation needs, with the supports and tools in place to do so. Some students, whether at the undergraduate or graduate level, choose to disclose their disability publicly in order to promote greater inclusivity and better accommodation for themselves and others. The decision to go public is highly personal. Using disclosure as part of a process of advocacy can be enacted in any number of ways. For example, students have engaged in public lectures for other students, faculty, or staff about their disability and educational experiences to enhance awareness, promote acceptance, or improve accommodation for that particular impairment or for those with disabilities in general; some have participated in committees advocating for greater accessibility or better access to accommodations on campus or in the broader community; and some students have worked closely with DSOs to advance knowledge about particular disabilities or effective strategies for accommodation, or have lobbied specific institutions, faculties, or programs for implementation of accommodations, etc. Such advocacy is often undertaken for altruistic or political motives with the goal of improving the educational experience for themselves, and for other students who may have similar or other disabilities. Such disclosure allows the student advocate to apply personal knowledge of discriminatory gaps to lobby for systemic improvements. The challenges inherent in disclosing one's disability in order to educate and advocate for more equitable treatment is often chosen by those students who find that the difficulties they encounter are balanced or even surpassed by the positive personal outcomes resulting from advocating.

What students should know prior to disclosure

There are many things students must ensure they have knowledge of before making the decision to disclose. They should have a general idea of how their disability impacts them in different educational settings, and what their needs are in these environments. For example, if they are in a lab setting, do they need help to perform experiments, or do they need help writing or editing reports? If they are in a graduate environment where they must conduct research and write a thesis, do they need a research assistant or an editor? The accommodations the students can avail themselves to varies greatly and may change depending on where they are in their program and may even change from class to class. They must know whom they can disclose to and be informed about their choice to do so. For example, do they strictly disclose to the DSO or do they work out accommodations with individual professors or within their peer support group? This process starts with open communication, which helps the student to feel comfortable and supported in making the decision. Students must also take into account their program of study, its requirements, and the potential for interaction between their disabilities and the learning environments in the program. As discussed above the process of disclosure is highly individual. Students and institutions must work together to ensure that students receive the information they need to make an informed decision.

Timing of disclosure

As with the choice to disclose, the time (point in the program, part of the semester) at which the student chooses to disclose is also highly personal. The student can choose to disclose at any point during his or her program, or not at all. Students must only disclose when they feel prepared to; this is particularly true for students who have invisible disabilities. Students with visible disabilities do not necessarily have to formally disclose by registering with their DSO. However, it should be noted that while students are free to disclose only when they feel prepared to, the length of time needed to design, develop, and implement appropriate accommodations—particularly for students in the experimental sciences, where accommodation solutions can be novel and/or custom-designed—also needs to be considered. A student who has identified an accommodation need that may take months to implement is better served disclosing that need well in advance of the time when that accommodation is required.

Students who have a known diagnosis and who may have received accommodations in high school or in the workforce may choose to disclose at the beginning of their program to ensure their continued success. Students who may not necessarily know about their disability, or who may fear stigma and being treated differently by peers and faculty, may make the decision to wait to disclose until they are not able to continue without formalized accommodations. Some students are able to complete their program without any formalized accommodations. Students who make the decision not to disclose may find they have to if a crisis occurs. If a student ends up having a rough patch and is no longer able to function without accommodations, he or she may be forced to disclose. In this case if a student wishes to continue with in the program he or she must disclose in order to ensure that accommodations can be put in place. Ultimately the student must make the decision to disclose at a time that best works for him or her.

A rubric for disclosure

The following list (adapted from The 411 on Disability Disclosure: A Workbook for Youth with Disabilities) can be used by students to help guide the disclosure conversation with both the DSO and other parties involved:

- Provide general information about the specific disability/ies requiring accommodation;
- Explain why the student has chosen to disclose his or her disability (e.g., its impact on academic performance);
- The type of academic accommodations that have been effective in the past (in undergraduate courses);
- The type of academic accommodations a student anticipates needing in the graduate environment;
- How a student's disability can affect his or her course of study.

Not everything on this list has to be used when disclosing; it is simply meant to offer some broad guidelines to aid in the process. There are many different approaches to how, to whom, and when to disclose. Within the postsecondary environment, it is important to put a plan in place for appropriate and accurate disclosure, if you choose to do so. No one form of disclosure works for all students, but it is important to understand various aspects of disclosure. With an increased number of students with diverse needs, understanding disclosure is critical. Given the sensitivity of this issue, it is important for different university departments to work collaboratively to create an inclusive postsecondary education environment.

Identifying the right players

In order to make an informed decision about disclosing, students must know who they can disclose to. When the decision to disclose is made, the student can choose to disclose to any number of people involved in his or her education. While the DSO is responsible for providing formalized accommodations the student must disclose to receive them. The student can also choose to disclose to individual professors or departments and work out accommodations on an as-needed basis. Trusted professors or departmental staff can help the student navigate the system and feel comfortable disclosing his or her needs. The person(s) the student chooses to disclose to is also heavily dependent on the specific disability and the program.

If the student makes the decision to disclose, deciding on to whom to disclose is an important step. Disclosure can be to the DSO, supervisor, other faculty members, department, etc. The student may have to disclose his or her accommodation needs to several individuals (faculty, department heads, etc.) before the accommodations can be discussed by a team.

The right player for disclosure is also heavily dependent on the situation in which the student is disclosing. For example, for overarching accommodations the student must disclose to the DSO but for classroom or program accommodations he or she must only disclose what pertains to that situation (e.g., to ensure they are not penalized if attendance is required and his or her disability prevents them from doing so on a regular basis).

Impact of disability types on disclosure

As briefly discussed above, disclosing can be heavily dependent on disability type. Students with various disabilities can be perceived differently, especially within different programs. It is important for students with visible and invisible disabilities to understand the benefits and drawbacks to disclosure, and how the choice to disclose (or not) can impact their academic career. Disclosure of invisible disabilities poses unique implications for students (Johnson, 2006) in the form of labels that carry significant stereotypes and societal stigmatization. Students often fear that by disclosing they will be seen as different from their peers and less competent. Students with visible disabilities must also choose whether to disclose during the application process or during the first meeting with their professors, potential supervisor(s), advisor(s), or departments. All students with disabilities face the questions of how much to disclose and disclosing a need for accommodation rather than their disability.

Pros and cons of disclosure

There are pros and cons to disclosure. Especially within certain programs, students may be concerned about the impact on their credibility that disclosure may have. The following highlights some of the benefits to proactive disclosure in one's academic journey. These are applicable across disciplines. Disclosure:

- Promotes openness and trust;
- Facilitates accommodation;
- Educates the supervisor or faculty advisor about accommodation;
- Sets out an expectation of fair treatment;
- Promotes open and honest communication;
- Protects against risk of discrimination. (Roberts, 2008)

Students, for a variety of reasons, may choose not to disclose. The following are some of these reasons:

- Risk of stigma;
- Loss of privacy;
- Loss of control of personal information;

- Reactive rather than proactive;
- No knowledge or understanding of disability or needs;
- No accommodation;
- Might not be necessary—work may not require accommodations.

As noted, disclosure is a highly personalized choice and one that is highly individualized. Students should be given the information they need to make an informed decision as to whether or not to disclose.

Implications of disclosure in laboratory and practical space environments

Students in the sciences need to consider the impact of disclosure of disability and/or accommodation needs in terms of the number of individuals they may need to talk to, as well as the impact of those conversations. Because of the complexity of the number and type of learning environments in the sciences—and also the fact that in a research lab environment or placement, or in fieldwork, a student may work with multiple individuals who are in teaching or instructional roles (many of them informally)—the student may be in a position of having to disclose (or choosing to disclose) to more individuals outside of the formal student–faculty or student–DSO relationship. Indeed, the student may choose these avenues of disclosure over the more formal approaches we discussed earlier in this chapter, a choice that has its own implications in whether or not the people the student works with are appropriately trained in disability sensitivity and response to disclosure.

As discussed in chapter 3, Barriers faced by students with disabilities in science laboratory and practical space settings, there are a number of significant attitudinal barriers individuals in the sciences may have about the participation of persons with disabilities in STEM. These attitudinal barriers, if not properly countered or educated against, will come into play when students choose to disclose an accommodation need. Faculty may try to play "gatekeeper," for example, or overstep their boundaries with respect to disability management and accommodations provision. Colleagues and peers, placed in instructional positions, may choose to socially ostracize or shun a peer with a disability. Members of the teaching team may prove recalcitrant in providing accommodations or assistance to the student. The student may feel he or she is suddenly being treated differently by peers, faculty, and the department, and may experience isolation and impostor syndrome as a result.

Other factors affecting disclosure

As briefly touched on throughout this chapter, the program and level of study a person is in can also greatly affect the decision surrounding disclosure. It can affect everything from the decision to disclose, to how much to disclose, and to whom. For example, students in STEM-based programs may be less likely to fully disclose due to the competitive and practical nature of these programs. Within these programs he or she may often be the only student with a disability within the program and fear becoming an outsider with disclosure. There are also various people to whom students may have to disclose.

When students enter graduate studies they are often unprepared for how different graduate education is compared to undergraduate studies. First, graduate education comes in several different flavors. "Research-stream" graduate programs at the master's and doctoral level primarily consist of a research-oriented thesis project and a few courses. "Professional" graduate programs are often short, intensive, theory- and practicum-based programs that act as entry points or necessary touchstones in certain regulated professions (e.g., occupational/physical therapy). There are also intensive course-based graduate programs, which may or may not require some research (though not at the level of a full thesis). Only this last type of graduate program will resemble an undergraduate degree. Overall, graduate education involves a greater degree of independent work/research; independent fieldwork, lab work, or practicum placements; less course work; a more significant degree of "learning outside the classroom;" a greater focus on research and thesis requirements; and, an expectation to publish and present academic work. Graduate research in general is a more student-driven learning process. Because of the nature of graduate studies, students lead their disclosure process, knowing who to talk to and what they should disclose. As opposed to undergraduate education, where the DSO acts as the go between students and faculty/departments, in graduate education, the onus is on the student to undertake this process. Students must take a more active role in implementing accommodations because the "standard" accommodations (i.e., those routinely offered by the DSO for undergraduate students) may not always be relevant. For example, extra time on a written exam is not applicable in the context of an oral thesis defense and assigning a notetaker in class is not appropriate for departmental seminars that may not occur regularly. Being an active participant in the accommodation process requires knowledge of the policies and practices of accommodation and/or disclosure and knowledge of the graduate administration in your department as well as other people outside of the DSO who may be important. This information may be readily available through departmental resources such as websites, calendars, and registration sites and will foster early disclosure and more prompt, effective accommodation. As part of the decision-making around disclosure, it is important to do this research upfront.

The student's relationship with his or her supervisor can greatly influence the success of graduate education. Students must decide what is best for them in terms of disclosure, based on their interaction with the supervisor, their own comfort level, their needs, and the other supports available to them. Some students may fear that their relationships with their supervisor and peers will change if they disclose.

Conclusion

Within this chapter, the process of disclosure, accommodation, and how it interplays with disability has been explored. We examined some common reasons why students opt not to disclose, and some of the pros and cons to disclosing. Although we outlined a process for disclosure, each student's needs are unique and individual and thus disclosure looks different for everyone. Students must take into consideration many factors prior to disclosure, but ultimately the decision to disclose rests with the student. There are many pros and cons to disclosure that must also be taken into account. Ultimately, the student only has to disclose his or her diagnosis to the DSO and accommodation needs to other parties. With an increased number of students with diverse needs, understanding disclosure is critical. Given the sensitivity of this issue, it is important for different university departments to work collaboratively to create an inclusive STEM education environment.

Student as ACTor—recognizing the importance of advocacy, communication, and trailblazing to student success in STEM

Chapter Outline

Introduction 85
Student as advocate 87
Student as communicator 88
Student as trailblazer 89
Steps for a successful secondary school to postsecondary transition 90
Conclusion 91

Dayna's Story: As an undergraduate student with a disability, I can attest to the excitement you, a fellow student, feel and the nerves you must be experiencing, as you contemplate the future. Instead of studying a technical field, I chose to focus my efforts in Communications and Science, Technology and Society. The long and short explanation of my degree is that it helps me understand and communicate the relationships entwining scientific and technological disciplines with society at large.

Part of being a communicator, and a person with a university degree, is acknowledging your shortcomings. I think I learned more in University about what I don't know than I did about what I studied. Part of the genius for me in finding my interest in Communications was that there was a world out there that existed within technology and science where I could be fascinated by the material and want to understand the concepts without actually doing any of the real roll up your sleeves, down and dirty, "hands-on" science. Once I found this Mecca, it was fairly easy for me to decide that this is exactly what I wanted to do. I wanted to be the type of person that explains technologies to people, so that my grandfather, a man with an eighth grade education from 1930, could understand what they were and why they would be useful. More than that, I wanted to do something that I could see was changing the world. Many of us, I think, have that dream and work to find ways to realize it.

I love books about pioneers: Every story, true or otherwise, about the first Europeans to see the Rocky Mountains or the Franklin

Creating a Culture of Accessibility in the Sciences. DOI: http://dx.doi.org/10.1016/B978-0-12-804037-9.00007-3

expedition which tried to chart the Northwest passage. History recognizes these people as Pioneers. Today we have Social Pioneers, people willing to chart a new course because they care enough about something to change people, to make them listen.

As a Communicator, it is most important for me that when someone needs to be heard I am there to listen. There are too many people in this world who do not feel heard. Under these circumstances they are compelled to compete for the opportunity to share their thoughts, instead of listening to other peoples opinions. This form of impatience inspires an endless cycle and creates one of today's biggest communications challenges; people are so afraid of being misunderstood that they refuse to understand anyone else. I see this a lot in communications around disability in university and college, and I know it happens frequently, especially in the sciences. We become so focused on getting our own points across that we forget to stop and listen to what the other person is telling us.

For me, clarifying miscommunication is not only a sign of respect for the person I am talking to but it is a method of making sure that my needs are understood and that the necessary accommodations are communicated to the professor, and my fellow classmates taken into account, so that I could be accommodated without hindering any of the other students. During my time at University some of my classmates have been willing to bend over backwards to help me, conversely I have some Professors who are offended at the very thought of making accommodations.

Mahadeo's Story: Nobody asked me to be a trailblazer or pioneer. Nobody told me I was going to be one, and nobody gave me any preparation for it. Yet, now, the identity of being the first blind cancer researcher has become deeply ingrained. It's the easiest way of explaining all that I have done to the world.

How did I get from there to here?

How did this identity come to define my career for better or worse over the last twenty years?

First of all, I think it's because it was true. I really am the first blind cancer researcher. To some degree, it doesn't matter if we're the first in the world, or the first in our class, department or school – what matters is that, in the experience of the people we have to advocate to and communicate with, there has been no one before us. To them, we are the first. To them, we are the pioneers, whether we like it or not.

Second, although it really is nice to be second sometimes – it would be lots less stress, I think – we can do only two things with the knowledge that we happen to be the first. We can reject that fact, and choose to not engage with others to understand what "being first" means to them, theoretically, practically or operationally. Or we can embrace it, run with it, and make it a part of us.

I tried both approaches – the first in my Master's, the second in my Doctoral program.

From first hand experience, I had challenges with rejecting the identity – or, if not outright turning away from it, at least shunning it for a couple of years. In fact, that method failed. It's possible that there were confounding factors in my Master's – and there were lots that could've impacted this issue – but the challenge is that we only have one shot at life, and very rarely do we get through chance, circumstance or effort a do-over. So, understand that my experience was not positive when I tried to shun being the first.

When it came time to do a PhD, I made the opposite choice – I embraced it.

I became an ACTor – an advocate, a communicator, a trailblazer – as opposed to one being acted upon.

It worked, in ways that other choice never did for me.

Introduction

Congratulations! And welcome to university. As a student with a disability, it is no small feat that you have made it this far. If you have found your passion in the fields of science, technology, engineering, and/or mathematics then perhaps you are looking to go a little further. Your excitement over your chosen field has brought you this far, but you will need to stoke that fire if you hope to achieve your end goal.

Throughout this book, we present an analysis of the barriers you face, and the solutions that you, your faculty, educators, and accommodation specialists may employ to lessen their impact if not remove them completely. In chapter 6: Disclosure in the sciences, we focused on the issue of disclosure of accommodation need in the academic context, and the challenges and considerations students with disabilities in STEM programs and disciplines need to think through. In this chapter, we expand on that discussion and focus very specifically on the roles you can play in being part of your own solutions. For faculty, educators, and service providers, this chapter offers insight into the student way of thought with respect to advocacy, both personal and systemic, and the potential impact of that advocacy in the engagement of students with disabilities in STEM fields. For students, this is one approach we think you can take in order to have some influence and control over your participation in the sciences during a very critical time of your life, education, and professional and personal growth. Of note, disclosure of accommodation needs is a prerequisite of the dialogue you, as a student with a disability, may have as part of the larger roles suggested by this chapter.

There are three words this chapter will identify you with. You may choose to agree with them or not, embrace them or not. They are not meant to define you but more to help you understand the part you have chosen to play. It is uncommon for students with permanent disabilities to get an education in the sciences, because these fields are seen, whether rightly or not (you must decide for yourself), as being difficult to accommodate, and being "impossible" for students with disabilities to succeed in (see the discussion of barriers that we presented in chapter 3: Barriers faced by students with disabilities in science laboratory and practical space settings). It is for this reason, apart from everything else, that this book was written.

It is important for you understand a philosophy upon which this chapter was written. "Impossible" is a word used far too often by those who work with students with disabilities in the educational context—"impossible" is often a euphemism for lack of knowledge, training, experience, or willingness to innovate. As students with disabilities, we have been told that because of our differences and the cost and effort it would take to accommodate us, it is far easier to simply not.

While how to accommodate students in the sciences is still in many ways uncharted territory, there are some things faculty members and schools can do to help students succeed in their programs. In essence faculty members can also be actors of a different ilk. This will be covered in more detail later in the chapter.

The three words that can be used to describe you students who venture into STEM fields are advocate, communicator, and trailblazer. In this chapter you will summarily be referred to as "ACTors." If you are feeling a touch of anxiety about being called any one of these things you need to remember the following:

- What is right for someone else might not be right for you.
- Everyone speaks, the key is hearing and understanding others so that you yourself can in turn be heard and understood.
- Just because you are one of the first to travel down a path does not mean you walk it alone.

You need to understand that being a member of the permanently disabled community means that you are bound (whether you know it or not) by your impression of yourself, as much as by others' impressions of you. When you understand enough about your own needs to articulate them plainly, and in a form that others can understand, you will see that communication and reception are two of the biggest challenges faced by students with disabilities. Whether or not you think you know what you need, articulating it to others and making sure that they do not find accommodating you to be a burden will be one of the most challenging things you will face throughout your educational journey.

These challenges and how you face them are the primary burdens for people with disabilities and above all other things are what will test you in your academic career and beyond. By acknowledging and making use of these challenges and the things you learn from facing them, you will attain strong control over your life. These challenges are a benefit of disability. They make us, maybe not better people but certainly better communicators, better listeners. By harnessing challenges and taking advantage of them you can control not just your own life and environment but you can influence other people's opinions of you and therefore their willingness to help you. Once you have found the right mentor, the right position, and the right field of study you will find yourself very in tune with your research and more than that you will be fascinated and challenged in ways that have absolutely nothing to do with your disability.

That is in fact the goal of "ACTing" and of accommodation as a whole. The purpose is to create an environment of seamless integration wherein students can feel they are no different from any of their peers, and their peers similarly are not impacted in any special way by them. While anyone can tell you we are leagues away from this form of seamless integration, every situation is a work in progress and improvements are always possible and can be made at any point during the academic year and your career by speaking with your disability services department.

This chapter identifies you the student as an "ACTor," someone who advocates, communicates, and blazes the trail for students with disabilities and workers who venture into the STEM fields behind you. While you may not be the first person in your field globally, nationally, locally, or even in your school, you are very likely to be the first person in your discipline to need your precise accommodations. The rationale behind this is very simple. There is no one person who is impacted by their disability in all the same ways as you. While someone's impact may be similar to yours in some situations, disabilities impact people based on factors like when they became disabled, how they became disabled, and the reactions of the people around them as a result of their diagnosis. This is where your school can help you. Faculties have the ability to empower students by taking on their own form of actor role.

Remember that you are a pioneer, and what you do as an actor in your own self-interest today could impact those who follow you for years to come. If this makes you anxious, it means you don't want to let anybody down. But remember—the only person you can let down is yourself, everything else is out of your control. There are no stories about pioneers who set out on a journey through unknown territory all on their own. There will be horns blowing, banners flying, and lights going off everywhere directing you to the help you need. The most important part of accessing accommodations is that YOU and only you are responsible for accessing them. If you choose not to identify yourself to the appropriate offices or if you are unwilling to receive help no one is going to hold your hand or force it on you. Even if you choose to disclose later on in your degree, you will not be alone.

Student as advocate

The first step in acting is being an advocate (see the previous discussions on self-advocacy in chapter 3: Barriers faced by students with disabilities in science laboratory and practical space settings and chapter 6: Disclosure in the sciences). Self-advocacy is something you must do for yourself. If you decide that your needs are such that something done even slightly differently would really help you then you must seek out that change for yourself. This makes you an advocate. Advocacy is important because, if done correctly, it may result in you getting the help you are looking for. However, if it is done incorrectly the results can be catastrophic and can impact far more than just you. Advocacy is often performed on behalf of the self with the secondary result of impacting someone else. It usually takes the form of knowledge and/or experience applied from your lived experience on your own behalf.

Being able to talk articulately on the subject (of your accommodation needs) and have yourself understood by somebody who is not well-versed in the subject matter at hand is crucial. Knowing the material you are talking about inside and out, whether it is potential accommodations or even different adaptations that would make things run more smoothly, can be tremendously helpful because you can show other people how simple accommodating can be. People tend to do things that don't require them to do a lot of work or spend a lot of money, that is just human nature. By ensuring them that helping you will not cause them more work you can get people on your side.

As stated earlier, what is right for someone else might not be right for you. It is important to be well versed in more than just the accommodations that you want to receive because knowing what these are and why they may or may not work for your particular needs will give you another edge in discussions with your school. The most important thing to remember when you are trying to collaborate to solve a situation with other people is to keep a cool head. There may be times when people refuse to accommodate you the way you feel you need. By keeping calm and making sure you are well versed in the material you can become your own best weapon.

While being an advocate means standing up and identifying a problem, it also means that learning to communicate even in difficult situations can be a true asset. Being a good communicator is something you must do for yourself. Being able to communicate politely and articulately to a variety of audiences with a constantly changing knowledge base is an asset. There is a good chance you will face antagonism, perhaps even hostility. This is especially true if your accessibility services department is newer or is not given much power within the school. Learning to handle these negative reactions is where advocacy, the ability to speak intelligently on a subject, meets communication.

Student as communicator

This may sound cliché but when communicating with anyone at any time, you will catch more flies with honey than with vinegar. In this metaphor the flies are the help you are asking for, the honey is you being nice, and the vinegar is you being anything else (sadly yes, sarcasm is included). A surprising number of people make this mistake. Just because accommodations are a right under the law in many countries does not mean in any way that the law can or will force any one person or group to accommodate you. By being nice to people and making them your friends you make them want to help you. If they want to help you so much more becomes possible. While it may sound manipulative, maintaining relationships with people who can help you is extremely important. By doing so you will keep open avenues to get help when you need it. Without this same opening you may metaphorically find yourself on a deserted island without a life raft.

It is also important to hear and understand others so that you yourself can be heard and understood. By listening closely to people and internalizing what they say, even if they are telling you why they cannot help, you might be able to use some of the information they provide you to your advantage. By taking their reasons back to your network (personal networks will be discussed in more detail in chapter 9: Peer-support networks) you might be able to use the information they provide you with as well as the resources within your network to come up with an alternative solution. By using your own network to research for yourself you can improve your chances of success. While other people may genuinely want to help you, you are not the most important part of their day. By staying at work on the issue you can be sure it moves forward.

Listening openly to people, without putting your guard up or becoming emotional, makes others feel important. So few people feel heard that when you actively listen and respond to what they are saying instead of waiting anxiously to say what you want to say, you will find people are very receptive to you and appreciative of that courtesy. This simple consideration of other people's thoughts and feelings will go a very long way toward getting them to help you. The act of listening is such a strength to the skilled communicator that when they find the people who can help them achieve their goals, those people are only too willing to go that extra bit to help them out.

By combining an aptitude for advocacy and a competence for communication we get the perfect combination of qualities to identify a trailblazer. This person has worked to hone his or her skills, or is naturally talented, and as a result is persistent and creative enough to get the job done.

Student as trailblazer

The term "trailblazer" is synonymous with "pioneer." These are people who choose the path of most resistance, the ones that spend their time looking for the hardest puzzles or the densest parts of a forest so that they can charge through and come out the other side. While this challenge might not be as deliberately sought out as the metaphor suggests. it may at times feel as though this is exactly the kind of challenge you have taken on.

Within a university there are set departments to help disabled students achieve their goals. These departments are not always appraised of all possible accommodations for all fields. This is because they have only needed to know what accommodations were necessary for the students already enrolled in their courses or programs. This is not their fault, but as a student in a nontraditional field, at least for people with disabilities, creative solutions and patience must be applied. When dealing with advocacy and communication every student is essentially the same. The trailblazer trait is what separates the different faculties. Students in arts, education, and other similar fields have been accepted and accommodated within most schools long enough that it is generally fairly simple to accommodate them. However, the STEM fields are not as adjusted, and still need students to help set the standards. While these faculties may or may not be doing everything they can to help you achieve your goals being aware of what is available as a student needing specific accommodations, and figuring out how the university can get those accommodations economically, is the first step to blazing a trail for all students with disabilities entering into STEM disciplines as programs of study and potential career paths.

All of this to say that your support networks can really help you in trailblazing. Students who set out to do this with anything less than a burning passion may feel overwhelmed and underestimated. The department and especially the school has much less incentive to help you get the accommodations you need for you than you do to get the accommodations for yourself. By relying on your networks of friends and family you can discover strength and tenacity that you did not know you had. These and a certain amount of steadfast stubbornness are traits you will need if you want to see this through to the end no matter what.

You are definitely not stepping out on the path of least resistance. When you leave the comfort and safety of your home, your high school, or whereever else you may be starting from, you must know that you will face challenges on this new path. Nobody can predict the exact challenges you will face or the accommodations you are guaranteed to want, but it is very important that you start learning about what you may need.

Steps for a successful secondary school to postsecondary transition

Now that you are in postsecondary school the only person who is responsible for identifying your needs is you. If you don't disclose your accommodation needs you won't be accommodated. It can be very tempting, especially if you have an invisible disability, to pretend you don't need help. But the only person you are hurting when you do this is yourself. By denying yourself the accommodations you need, you do not give yourself the best chance of succeeding.

It is very important that you know your stuff. Do your best to get an idea of how your school structures their classes. What exactly is the difference between a laboratory and a lecture? A tutorial? A seminar? What can your disability services department do to help you? What kinds of accommodations are available through them? These are just some of the questions you will need to find answers to. While you do not need to know everything about your needs at the beginning if you can formulate a working strategy for yourself you will find your life is much easier. You can tweak your accommodations as you go if they are not working.

One of your most valuable resources and often the one most underestimated is the people around you. By knowing their talents, passions, how they would be willing to help you, and if they cannot, who they know who might be willing to help can be a huge boost. It is also important for you to understand what help you think you will need and how you want people to help you. Your connections are important but people may want to provide more or less help or than you actually need. This is discussed in more detail in chapter 9: Peer-support networks.

Knowing what services your school currently offers and which ones are available to people with your disability can also be very helpful. Knowing what can and cannot be made available on short notice or with minimum effort by the support staff is important as well. This information can help you when you are in a jam or find yourself unable to move forward because of some conflict.

Conclusion

It is not uncommon for schools, facilities, or individual professors to be resistant to helping students who have requested accommodations. At times, people spend more time and energy fighting an accommodation than they would spend simply implementing it.

The possibilities for how to accommodate people are endless and for smaller institutions and those who have less than the support they need, accommodations can seem like a burden. Most accommodations take only a small amount of flexibility and some systemic changes to implement (e.g., things like letting students take longer to write exams are relatively easy to implement). By implementing them your faculty is improving your chances for success. But more than that, if a student is the first in his or her field that you are aware of, even taking enough of an interest to Google and find out about other people in his or her field might give you a sense of what the student might need help accommodating or facing.

While all of this is important, the single best thing any faculty can do is to ask a student for information about what he or she needs. By reaching out to the student you are acknowledging his or her needs as a member of the institution and also giving the student a point of contact for anything he or she needs in the future. This does not mean that all faculty should have their own accommodations person. Quite the opposite in fact. The proposed meeting is best held between a member of the faculty who has the ability to make decisions and implement action on behalf of the student, a member of the disability services office who is the student's main point of contact there, and the student. This dream team can be the most efficient at implementing unusual accommodations, particularity in labs, but none of it will happen spontaneously.

While the student is responsible for setting up this meeting, the faculty should be open to this type of collaboration and come prepared to help. If this is the first student with a disability to attempt this field at this school it is very important that the faculty play their part, but so the student. Once these lines of communication have been established they must be maintained for the duration of the student's education.

Mental health and well-being for students with disabilities in the sciences

Chapter Outline

Introduction   95
The stress of being a trailblazer   96
Impostor syndrome   97
Stress of disclosure, accommodation, and disability management   99
Disability and the stresses of STEM training   100
Recognizing signs of stress   101
Strategies to improve well-being and maintain balance   102
Conclusion   103

When I started my Master's, I was the first person with a visual disability in my department. I'd been recruited to the lab at least in part because of my background in genetics and molecular biology – and the importance of molecular techniques in disciplines outside molecular genetics and biochemistry was just beginning to rise. Today, of course, these techniques are entrenched in a variety of allied disciplines, but in those days, I was there to bring something new to the lab group, and even the department.

Me.

Just out of my undergrad. Almost no research experience to speak of.

Four years younger than my cohort. Supposedly immature.

And – were I to believe the naysayers – someone with a visual impairment, who had no business whatsoever being there.

I was undaunted.

Until the very first thing went wrong. And then the second. And the third.

Then I had doubts. Who was I to think I was better than my peers? Even with the accommodations that I had, but definitely without them. Did I really know what I was doing? Did I deserve to be in graduate school? Did I deserve to be in that lab? Did I even know what it was like to be a scientist? Should I stay on this path?

Then things got really bad – everything started to go wrong.

It was like a tailspin – nothing I tried allowed me to get out of it. No experiment worked.

Creating a Culture of Accessibility in the Sciences. DOI: http://dx.doi.org/10.1016/B978-0-12-804037-9.00008-5

My assistant was working four days a week, 30 hours total. I was working five or six days a week, going in on weekends, and pulling 10, 12, 14 hour days. I was – literally – waking up, getting dressed, leaving home, going to the lab, banging my head against a figurative wall, leaving the lab, going home, going to bed. Nibbling on food somewhere in there.

And repeat.

Over and over again.

For months.

Needless to say, that didn't work. Never mind very well, that didn't work at all. For months, I literally cut myself off from my support network. I went through the motions, without anything really registering with me. I gave up tutoring, an activity that I found very fulfilling. I let go of most of my volunteer commitments, and didn't follow through on much of what I had left. I won awards – academically and in the community – during this time, but everything felt hollow and meaningless. I was, if anything, even more introverted than usual.

Metaphorically, I was drowning.

And clinging to the one thing that was making me drown.

Not one of my proudest moments – indeed, one of the worst times of my life. I didn't realize it at the time – and didn't have the vocabulary or willingness to articulate until much later – but I had significantly and negatively impacted my well-being and mental health. And, in attempting to force my way through that wall, largely made things worse, not better.

It took eight months to solve the technical challenge that started the spiral – well, not so much solve, as totally work around it. It wasn't until two or three years later that I figured out how to deal with the original technical problem. (You know what? It was a pretty simple, "Why didn't I think of that?" technical fix too.) But it took a lot longer than that to deal with the psychological consequences – it didn't help that through my PhD and into my postdoc, there were a few more similar hammer blows to my psyche. What did help those times was that I did learn one lesson, and that was how to maintain my balance and well-being. I managed to find things – volunteering, giving back to my community, cooking, spending time with friends, walking – that served the purpose of getting me physically and mentally out of the lab, and into a different headspace and environment. Put another way – I found things I took joy in and managed to hold on to them through any challenges I faced in the lab, or in my training.

They say that once you tear a ligament or a muscle, it's a lot easier to tear it again, unless you keep up your exercises and due diligence. The same, I think, is true of our well-being. Once damaged significantly, it is very – surprisingly – easy for it to happen again, particularly when you least expect it. Even on the cusp of what ought to be one's greatest successes.

Such is the case when you are – whether you like it or not – a trailblazer. Whoever said it's easier to come second was onto something, I think – except when we have no choice in the matter.

Such is the case when you're in a field where systemic attitudinal barriers are significant – except from those who know you directly and can see what you're capable of for themselves.

Introduction

STEM training is a high-pressure environment. Undergraduate students feel pressure to succeed academically as well as in the extracurricular environment in order to have the best chance at successful entry into a desired professional program (e.g., medicine, dentistry, or law). Students who choose to pursue graduate education feel a different sort of pressure, associated with the culture of academic training and the requirement for productivity in the form of research and publication in graduate and postdoctoral training environments. All students feel pressure to compete against their peers and succeed in the labor market. To be sure, these pressures are not unique to STEM, but they are most visible and prevalent in these fields.

Stressors that students may face in STEM training include, but are not limited to, the effort involved in establishing a professional identity and brand; attempting to achieve a "school-work-community-life" balance in postsecondary education; the stress of competition with one's peers for grades, scholarships, etc.; the effort invested in choosing one's career path and launching one's career; the effort invested in self-promotion; the time needed for building and maintaining relationships with faculty members—particularly, in graduate school, with one's thesis advisory committee and thesis supervisor; coping with perfectionism; interacting with one's peers; getting adequate sleep; and dealing with the culture of postsecondary education—particularly, again, at the graduate level, with its demanding research culture and the environment of graduate school.

Life stressors may also play a role in the student's mental well-being, as it is often very difficult to compartmentalize and separate academic from nonacademic stress in the student's thoughts. The student may have to deal with the stress of being on his or her own, away from family, friends, or a support network; wrapped up in this is the potential for culture shock if the student has come to a new city or country. The student may deal with stress from family situations and/or social isolation as well as from other life events.

Students with disabilities entering and proceeding through STEM training have an additional series of stresses to cope with that can lead to an increased feeling of pressure and greater potential for decreased mental well-being. Specifically, students with disabilities may be prone to "impostor" syndrome—a feeling of inadequacy or of not belonging in STEM. Students may feel significant pressure because of the sense that they are alone as persons with disabilities in their fields, departments, or institutions. Related to that, students may feel stress from being trailblazers.

In this chapter, we look at some of the major factors that can impact a student's mental health and well-being, discuss ways that students, faculty, and disability services staff can recognize some of the signs of poor mental well-being, and conclude by discussing the importance of balance and mental well-being to students with disabilities in STEM.

The stress of being a trailblazer

We have acknowledged previously that persons with disabilities are significantly under-represented in the sciences, in stark contrast to the general population statistics (see chapter 1: The landscape for students with disabilities in the sciences). It may well be, then, that a student may be the only person with a disability in his or her research group, department, or field. We have discussed previously the advantages of being in such a position (see chapter 4: Student perspectives on disability—impact on education, career path, and accommodation), but let us now consider the potential impact of being a trailblazer on a student's well-being and mental health.

There is no one to relate to. A student in the position of being "first" may feel a sense of social isolation. After all, no one else (to his or her knowledge, and the knowledge of the faculty and disability services staff) has had similar experiences, or felt the same things. Unless the student is able to reach out beyond his or her discipline, or beyond the institution, or even across national borders, he or she may never encounter another individual with a disability in the sciences in person. To be sure, other such individuals do exist, and countering that feeling of isolation is one rationale for writing this book, but until the student connects personally with someone else, that sense of isolation—of "no one can understand what I'm going through"—prevails.

There are no precedents. Many accommodation supports are based on prior experience—what has worked for the student in the past, what has worked for previous students in similar situations, prior knowledge of faculty members and disability services staff. However, often, a student with a disability in the sciences might be the first within the experience of faculty and service providers (see chapter 2: Accessibility and science, technology, engineering, and mathematics—the global perspective). Precedents for working with the student are currently limited at best, nonexistent at worst—which could lead faculty and professionals, no matter how well-intentioned, to suggest potential solutions that may be unworkable for the student. The student, themselves in unfamiliar territory, may not know enough, or have enough experience, to counter with ideas of his or her own. Thus the student may end up in situations that are inappropriate, given his or her interests, disability, and accommodation needs. Having to navigate through those situations, and potentially find solutions to challenges, when everyone around the student professes some level of professional inadequacy, adds to the stress.

There are many systemic attitudinal barriers. As discussed at length in chapter 2, Accessibility and science, technology, engineering, and mathematics—the global perspective, many faculty and educators take the perspective that students with disabilities are not suited for work in the sciences, for a variety of reasons. This attitudinal barrier might impact faculty opinions in admissions committees, in class, and in the research environment. Beyond the education environment, these attitudinal barriers can be found in the scientific workplace as well. Constant exposure of the student to what amounts to indifference at best, and active negativity at worst, becomes very draining, exacerbating the sense of isolation discussed above, and contributing to an absence of physical and mental well-being.

Low expectations. In many ways, being confronted with zero expectations, and an environment where almost any strategy, no matter how successful, is considered positive, would actually be better for the student, than an overt expectation of failure. Students with disabilities in STEM fields often find themselves in situations where, not only is their success not expected, but their failure is. It becomes very difficult for any student to not internalize the sense, passed along from others around them, that he or she is destined to fail in the field. We will revisit this momentarily, as it acts as a contributor to impostor syndrome, another class of stressor that a student with a disability in STEM may experience.

Impostor syndrome

Impostor syndrome is defined as a "psychological syndrome or pattern. It is based on intense, secret feelings of fraudulence in the face of success and achievement. If you suffer from the impostor phenomenon, you believe that you don't deserve your success; you're a phony who has somehow 'gotten away with it.'" (Harvey, 1984). Impostor syndrome is common within the academic environment, particularly at the graduate student level—in STEM fields, particularly, where productivity is a significant measure of a student's success.

For students with disabilities, impostor syndrome is a clear and present challenge. This may be due to several factors: Students may be confronted with peers and faculty who may harbor low expectation for success; students may find themselves in an environment where they are not provided accommodations to enable their full participation in STEM; students may experience an absence of peers or role models with disability; and, students may already be exposed to the systemic perception that they do not "belong" in their program, discipline, or field. Impostor syndrome can arise on two fronts—when a student with a disability has been able to demonstrate his or her success ("my success doesn't belong to me; I don't deserve it"); or when confronted with failure or frustration when things don't work in the lab or during his or her training ("everyone is right; I don't belong here").

Impostor syndrome presents itself as a series of feelings or thoughts in the mind of the student:

- Feelings of phoniness and self-doubt (*"I am not as smart as they think." "I'm not as smart as I think."*)
- Fear of being "found out" (*"It's only a matter of time before people realize I don't belong here." "People were right—I don't belong here. It's only a matter of time before they kick me out."*)
- Difficulty taking credit for one's accomplishments (*"I don't deserve to win this award."*)

- Frustration with inability to meet self-set standards (*"I'll never be as good as I want to be, so why bother trying?"*)
- Lack of confidence, fear of making mistakes (*"I don't think I have what it takes to be a scientist."*)

(List adapted from the University of Waterloo Centre for Teaching Excellence: https://uwaterloo.ca/centre-for-teaching-excellence/teaching-resources/teaching-tips/ planning-courses/tips-teaching-assistants/impostor-phenomenon-and)

Several external signs of impostor syndrome in a student with a disability may manifest themselves: The student may become less accessible to his or her peers; be unable to motivate or respond to students or mentees; be uncomfortable acting as a role model or mentor; exhibit decreased research productivity or effort; put less effort into his or her classes; be less willing to present in public or to publish his or her work; be unwilling to attend events in the department; and, be unwilling to socialize (University of Waterloo, Centre for Teaching Excellence).

Transition points (e.g., undergraduate to graduate student; master's to PhD; teaching assistant to novice instructor; PhD candidate to postdoctoral scholar; graduate student to expert in your field) in a student's training are monumental, and as each transition point brings with it new responsibilities, new environments, and potentially new accommodations as well as new or reinforced systemic attitudes (University of Waterloo, Centre for Teaching Excellence).

Several strategies have been reported to be effective at managing impostor syndrome in students generally. These include:

Break the silence. Speak out about your feelings. Knowing there is a name for these feelings and that other people experience them also can be very reassuring. As a student with a disability, knowing that your nondisabled peers may experience something similar helps with a sense of belonging.

Separate feelings from fact. Everyone feels stupid from time to time. Just because you feel it doesn't mean that you are.

Recognize when it's normal to feel fraudulent. When something is new to you, you may feel like you are underqualified to perform an action. These feelings are natural response for any novice.

Accentuate the positive. Don't obsess over everything negative. Do a great job when it is important, recognize when you've done so. Seek the positive in all situations.

Develop a new response to failure and mistake making. Learn from your mistakes and move on. Don't dwell on what has happened in the past, even if you think others will. What matters is that you demonstrate growth and progress in your field—everyone will make mistakes from time to time.

Right the rules. Don't feel like you always need to know the correct answer. Recognize that you have just as much right as the next person to make a mistake or ask for help.

Develop a new script. Rewrite your mental script from "I am an impostor" to "I may not know all the answers, but I am smart enough to figure it out." Or: "I have just as much right to be here as everyone else, I can do this."

Envision success. Instead of thinking of worst case scenarios, imagine yourself conducting an excellent presentation or answering questions with the correct reply. Imagine success in your field for yourself.

Reward yourself. Learn to pat yourself on the back when you deserve it. Don't hide from validation!

"Fake it 'til you make it." Take a chance and "wing it;" this is not a sign of ineptness or desperation, but rather a sign that you are intelligent and able to rise to a challenge, by adapting to new and changing circumstances.

(List adapted from "Impostor phenomenon and graduate students," University of Waterloo Centre for Teaching Excellence: https://uwaterloo.ca/centre-for-teaching-excellence/teaching-resources/teaching-tips/planning-courses/tips-teaching-assistants/impostor-phenomenon-and)

Stress of disclosure, accommodation, and disability management

We have already highlighted the importance of the decision to disclose (see chapter 5: Key role of education providers in communication with students and service providers), and we will revisit the importance of that decision in the dialog around accommodation in the latter half of the book. Given the complexity of the STEM training environment, the thought process around disclosure and accommodation is not as simple as "Do I tell my disability services staff to get accommodation in the classroom?"

Students with disabilities, at various points in their training, will have to consider whether or not to disclose to peers, in the form of group members, lab mates, or colleagues on fieldwork assignments; to their students and trainees, when they are in teaching and leadership roles; or, to their thesis supervisors, at any level of training. Students with disabilities must also factor into their thought process the notion that the disability services staff alone may not be able to be of much help, and other subject matter experts may need to be engaged. Although none of this will happen without the student's permission, it is important for faculty and service providers to understand the stress that this places on the student, whose natural instinct may often be to say as little as possible about his or her disability, to as few people as possible. This is particularly true for students with diagnosed psychiatric disabilities, for whom a self-advocacy-based model of accommodation and engagement in education has been reported to be ineffective.

We recognize the irony inherent in our advocacy for collaboration and cooperation as a way to break down the systemic barriers in STEM facing persons with disabilities—one of the central themes of this book—being itself a source of stress to students. However, it is worthwhile to point out that having a successful education experience requires the full participation of the student, irrespective of disability.

Another stress factor for students revolves around their disability management strategies in the context of their STEM training. In particular, persons with episodic, cyclical or progressive disabilities may have their learning and functioning in a STEM-related laboratory change over time. A student may have successfully evolved one set of disability management strategies in time for his or her learning and function in a lab to change yet again, and the need will arise to have to develop a new set of coping strategies. Furthermore, students with acquired disabilities (who are diagnosed or identified during their training periods) are going through an adjustment period while they are studying, and must cope with the learning and function differences associated with their disability. The student's management of these changing circumstances can be a source of stress, as this takes energy and effort on the part of the student. While the student may not choose to disclose the specifics of his or her disability or circumstances—and, indeed, the student may very much want to proceed through training without what he or she may perceive as "special treatment" to single them out from nondisabled peers—the effects of the student's management strategies may become evident as decreased well-being. While peers and faculty members may not know or understand the underlying reasons, they may be in a position to recognize the consequences of the student's stress. We will return to how best to recognize signs of decreased well-being and the potential for mental health issues a bit later in this chapter.

Disability and the stresses of STEM training

As we acknowledged in the introduction to this chapter, STEM training—particularly at the graduate level—is a high-pressure, high-stress environment. Indeed, the very nature of the training environment has been found to lead to mental health issues and diagnosis of psychiatric disabilities, resulting in the need for formalized supports for research trainees in STEM (Tsai & Muindi, 2016). In this section, we review the impact of disability on a student's ability to manage the stress inherent in the STEM training and work environments.

Undergraduate STEM training is often viewed as a means to an end—with secondary program options inclusive of professional faculties and graduate school. Success at gaining entry to these secondary programs is dependent upon academic success and merit during one's baccalaureate training. Furthermore, many undergraduate university and college STEM lab environments are, by definition, very rigorous and timed in their nature. Thus a student with a disability may experience very systemic stresses associated with the nature of STEM training overall, which are magnified in some respects through a disability lens—e.g., how does one establish a career identity for oneself in a field where the prevailing opinion is that you do not belong, and don't have the ability? The student may also experience very granular stresses associated with his or her course and labwork—for example, the need to finish an experiment within an allotted timeframe when various tasks take longer to accomplish due to one's disability.

At the graduate level, the issues of impostor syndrome, career direction, brand identity and disability management reflect on and compound one another—thus, stress brought on by systemic issues gets magnified, while a different set of granular issues come to the fore. The breadth of the graduate environment—from the lab to classes to teaching to seminars to conferences to publications to fieldwork to collaboration to the thesis defense to career and skill development—likely requires different coping strategies and accommodation in different settings. Coupled with the length of graduate training, the potential for evolution of the disability, research project and program, and the need to demonstrate research productivity as a measure of success, it is not uncommon for students to feel out of their depth and overwhelmed.

Particularly at times when milestones are to be reached (e.g., the thesis defense), feelings of inadequacy and depression are common.

Thus, taken together, given a series of systemic and granular, disability-related and environmental conditions, there is significant potential for the evolution of mental health issues and decreased well-being in students with disabilities in STEM. The next section will discuss how peers and faculty can recognize signs of stress in students with disabilities, and we will conclude with some suggestions for maintaining balance and well-being.

Recognizing signs of stress

A student who is experiencing decreased well-being due to stress will exhibit several signs in the lab or classroom, or through his or her behavior or physical appearance. In extreme cases, signs of stress may take the form of safety or emergency concerns. If, as a faculty member, peer or disability services staff person, you think that a student with a disability is exhibiting one or several of these signs, an open and non-confrontational conversation path is most appropriate. It is important to ask permission from the student to discuss your observations and perception that there may be a problem the student is encountering. Let the student tell his or her story—practice active listening, and give them your undivided attention. Acknowledge, be respectful of and support his or her courage in opening up and disclosing a personal difficulty. If you can, provide advice and suggestions, but do not mandate anything to the student. Emphasize that he or she does not have to deal with these issues alone; close the conversation positively and leave room for him or her to approach you for further discussion (adapted from the "More Feet on the Ground" website, Brock University, Canada: https://brock.morefeetontheground.ca).

Signs for all students, both nondisabled and disabled students under stress can include:

Classroom or lab indicators: Repeated absences; missed appointments; deterioration in productivity; disorganization; erratic performance; continual seeking of special provisions; pattern of perfectionism; disproportionate response to evaluation.

Behavioral indicators: Statements of distress; difficulty controlling emotions; anger, hostility, yelling, aggressiveness; more withdrawn or animated than usual; expressions of hopelessness or worthlessness; crying or tearfulness; severe anxiety or irritability; demanding or dependent behavior; unresponsiveness to outreach; shakiness, tremors, fidgeting or pacing.

Physical indicators: Deterioration in appearance or hygiene; excessive fatigue and exhaustion; changes in weight; statements about changes in appetite or sleep; noticeable cuts, bruises or burns; frequent or chronic illness; disorganized speech; rapid or slurred speech; confusion; unusual inability to make eye contact; arrival at the lab or in class bleary-eyed or smelling of alcohol.

Safety indicators: Written or oral statements that mention despair, suicide, or death; severe hopelessness, helplessness, depression, social isolation, and withdrawal;

statements to the effect that the student is "going away for a long time"; written or oral expressions of a desire to injure or kill someone else.

Emergency indicators: Student is physically or verbally aggressive toward her/himself, others, animals, or property; student is unresponsive to the external environment; s/he is: Incoherent or passed out, demonstrating a severe disturbance of cognitive, behavioral, or emotional functioning, or displaying disruptive behavior that appears to be out of control; the situation feels threatening or dangerous.

Other indicators: Expressions of concern about a student from his or her peers; a hunch or gut-level reaction that something is wrong.

It is worthwhile reiterating that the list of signs of stress described above may be found in anyone, disability or not, in the context of postsecondary education and during all levels of STEM training.

(List adapted from the "More Feet on the Ground" website, Brock University, Canada: https://brock.morefeetontheground.ca).

Strategies to improve well-being and maintain balance

Many college and university campuses have prepared mental health strategies and plans, in order to promote stress management and well-being among the student, faculty, and staff populations. While it is up to the individual student to determine what works best for them, most strategies involve taking time away from the lab or classroom setting—essentially, giving oneself a break from the pressures of STEM training as a student with a disability. Some students identified peer support (see chapter 7: Student as ACTor—recognizing the importance of advocacy, communication, and trailblazing to student success in STEM) as a coping strategy that works for them; others engage in physical activity (e.g., the gym or yoga) or team sports. Still others engage in social justice or community service and volunteerism, while some consider household chores (cooking, cleaning) to be therapeutic.

Much literature has also been devoted to the importance of a healthy lifestyle (nutrition, physical activity) on mental health and well-being, as well as to the importance of positive thinking—negative thinking reinforces poor well-being and mental health while positive thinking acts as a counter to those feelings. Building up one's resiliency and critical thinking skills can also be of benefit in managing well-being and mental health, as the student is then able to think his or her way through and articulate concerns or difficulties a lot better and more effectively. Cultivating positive emotions through approaches such as gratitude journals or activities that you find joy or happiness in doing is also considered of benefit. Maintaining a healthy and engaged social circle is crucial to this approach as well—minimize negative, counter-productive, or abusive relationships while expanding positive, supportive and encouraging ones. Balance and well-being are maintained and enhanced when individuals choose quality over quantity—a healthy social circle does not have to be large, but it does have to be integrated into your life in a meaningful and caring way.

From an academic or STEM training perspective, as well as in life, it is important to set realistic goals and expectations. Unrealistic goals, which cannot ever hope to be achieved, enhances negative feelings and poor mental health. If you feel comfortable talking about your well-being, it may be of value to challenge, with others, the assumptions you have about yourself as well as your worldview—consider whether you have internalized others' beliefs, or whether your own beliefs may be contributing to your mental health.

Consider the nature and extent of volunteerism or social justice or community work you may want to engage in, and what would be beneficial—also consider how formal (committees, clubs, boards) or informal ("doing your own thing") you wish to be with respect to giving of yourself or giving back to the community. A cautionary note: It is important to not over-commit to so many things, so deeply, that they supplant your education or career goals, or that they themselves become sources of stress and necessitate another cycle of seeking balance. Finally, consider expressing your spirituality. Spirituality involves the expression of self-awareness and self-acceptance, as well as connecting meaningfully with something beyond oneself, and developing a sense of meaning and purpose. As with volunteerism—which does much the same thing—this awareness enables a better sense of balance and perspective.

In situations where all these strategies fail, it is important to remember that mental health supports on campus and in the community are available, and it may be necessary to avail oneself of them. Indeed, talking through one's concerns with a trained therapist may be beneficial, even if employing other strategies to enhance well-being.

Maintaining mental health and well-being in the face of challenging and stressful environments, such as those found in STEM training, is ultimately a matter of balance and maintaining perspective. It is important to remember why we choose to engage in STEM training and what benefit we hope to derive from it, but it is important to put that in perspective and not to sacrifice well-being, balance or the people and activities we find joy in, in order to see that one goal to completion.

Conclusion

For a variety of reasons, the STEM training environment, particularly in graduate school, is a stressful, high-pressure one. Students with disabilities have to navigate additional stresses that may combine with those endemic to the learning environment and lead to decreased well-being and mental health. While we have painted a daunting picture of these stresses and the impacts they have on students with disabilities, it is important to remember that, with an appropriately supportive network and effective stress-management strategies, students can enjoy fruitful and rewarding training periods in the sciences. When recognizing warning signs of stress and decreased well-being in a student with a disability, faculty, disability services staff, peers, and members of his or her social circles should feel it is appropriate to engage the student in dialog to help them understand the challenges he or she is facing, and to help them find the appropriate coping and stress-management strategies.

Peer-support networks

Chapter Outline

Introduction 106
Why is peer support important? 106
What is peer support? 107
Types of peer support 108
Benefits of having a peer-support network 109
Potential challenges in having or maintaining a peer-support network 109
Operational issues: ensuring effective peer support 110
Peer support versus peer mentorship 110
Peer support: the beginning of the conversation 111
Where do we find our peers? 111
Virtual (long-distance) peer support 112
Peer support outside disability 112
Conclusion 113

I didn't start out with a peer support network.

I didn't even start out searching for a peer support network.

I was very much a loner during my high school days, and that continued during my undergraduate education and into my Master's. It wasn't that I was anti-social mind you – I liked to socialize fine – I was more asocial. I was entirely agnostic to the concept, it was never something I felt I was good at or saw much point in. After all, I was there to get good grades, not to interact with people, right?

There's an argument to be made that education is more and more an extrovert's enterprise, and that extroverted behavior in learning environments is being praised and fostered, while introverted behavior is more of an experience that fosters personal growth and evolution.

For us introverts, let's be honest, peer support groups are hard – they require a lot of opening up about our experiences and stories to people we don't necessarily know and may not feel comfortable around.

Not having a peer support group or framework in high school or undergrad or even my Master's meant that I'd chosen to go through my education on my own, and that I was independently trying to navigate any issues or barriers that arose, without any kind of sounding board to fall back on. This wasn't just a disability thing – this was an academic path issue as well, and in fact, that interface

Creating a Culture of Accessibility in the Sciences. DOI: http://dx.doi.org/10.1016/B978-0-12-804037-9.00009-7

between disability and academic path was crucial, and where I perhaps missed that peer support network the most.

Maybe "missed" is the wrong word.

"Wished I had" maybe?

When I started my PhD, I set out to fix this challenge, and I began to look for peer support frameworks. I started out with group supports – student groups on campus were the places to start for me – but over time evolved away from formal groups (even groups as a whole) toward more individualized peer support frameworks for myself. Individualized, one-on-one, peer support was more comfortable for me, I thought – I wasn't always comfortable in a group setting, and ultimately poor group dynamics drove me away from some of the groups I engaged with.

I felt like I needed very specific things that groups were not able to provide in the long run (although I did learn and contribute a lot as a member of those groups).

Today, my peer support frameworks literally come from my peers – usually coffee chats, usually one-on-one, and usually in the context of conversations about career and professional development. As my own needs have evolved, I have been able to better match my peer support network to those needs and interests.

Today, I highly value my peer support network, because today I meet that need on my own terms, and in ways that complement who I am.

Introduction

In this section, we have identified both the importance of disclosure of accommodation need, as well as the potential challenges and considerations a student with a disability studying the sciences may need to consider; we have expanded on the disclosure conversation to discuss the role of student as ACTor—advocate, communicator, and trailblazer—in the larger context of disability and the sciences; and we have highlighted the potential for negative impact on mental health and well-being. One significant coping strategy around positive well-being and homeostasis or balance in a stressful educational environment, like that of science education, is the presence of a peer support network. In this chapter, we will discuss the types of peer support networks that can be formed, and the benefits of peer support systems to students with disabilities studying in science, technology, engineering, and mathematics (STEM).

Why is peer support important?

In education, a student's mental and social health are just as important as, if not more important than, his or her ability to do the work required in the course of study.

One of the most detrimental experiences for students in programs and disciplines where disabilities are generally rare, like the sciences, is social isolation (see chapter 8: Mental health and well-being for students with disabilities in the sciences). Some of the ways to evolve a positive experience from this social isolation and the notion of being a trailblazer were discussed in detail in chapter 7, Student as ACTor—recognizing the importance of advocacy, communication, and trailblazing to student success in STEM, when we presented the ACTion model for student engagement. However, this problem also has a social element to it, as students who feel isolated can find school much more challenging. Disability and accommodation are already an isolating set of circumstances for students, but having a disability in the context of the sciences can take this isolation to an extreme.

Peer-support networks are one coping mechanism that can be used to manage this set of issues. The essential role of peer support is to provide a human connection, an individual or a group of people who are encountering the same or similar experiences and challenges. These people can perform many roles, from sounding board and sympathetic ear, to being individuals with whom to brainstorm potential solutions and from whom we can learn.

In chapter 7, Student as ACTor—recognizing the importance of advocacy, communication, and trailblazing to student success in STEM, we discussed using advocacy and communication to build beneficial relationships within your school and community. During that chapter we also discussed how a school's faculty and staff have the ability to be actors and champions on behalf of their students. Peer support is another component of this ACTion model. Remember when we discussed reaching out to networks of people who have your disability or one like it to get recommendations for accommodations as a starting point? Those people can be members of your peer-support network.

What is peer support?

While peer support is a broad arena and can be difficult to define, it can be comprehensibly represented in recipe card form (see Fig. 9.1).

From the ingredient list breakdown we can see that peer support covers many different areas. These people can be a release valve for things that you've kept bottled

Peer support	Instructions for use
3 parts friend 2 parts confidant 2 parts inspiration 1 part mentor 1 part lightning rod	Take one friend, find in them something you admire. Complement them on it. Outline a problem or concern to them. Ask their opinion on it. Carefully consider their opinion then repeat until satisfied with the conclusion you have come to. (Note: when outlining concern feel free to unload candidly, you'll feel better)

Figure 9.1 Recipe for peer support.

up inside or they can be a check-in for your mental health. They can also be friends and mentors, and can be any age, local or online. There can be a whole group of people that get together to discuss circumstances or it can be a one-on-one experience. Participants can be similar in age or can have a large generational gap. While each of these variables can change the relationship slightly it is most important that there is a mutual respect and understanding between the members. This respect and understanding breeds a form of peer support called peer mentorship.

Types of peer support

Peer support comes in a variety of forms, both formal and informal, in group settings or through individual interactions, through external facilitation or self-organization, and may occur through in-person or online. The type and quality of peer support may also be defined by what type(s) of support or interactions you, the student, are looking for—are you more interested in the experiences of fellow students with disabilities in your discipline, in the sciences generally, or in higher education? Or, are you interested in academic experiences in your field of interest, irrespective of disability?

Your school, faculty, or department may have a number of peer-support resources or groups already created, in the form of student associations, student clubs, special interest groups, or study groups. Looking beyond your individual institution, you may find special interest groups, committees, or trainee councils created by professional societies in your field or discipline. There are several professional societies and nongovernmental organizations that work in STEM and diversity broadly. A few of these may have specific interests in disability, high school, and/or postsecondary education. Additionally, you may find, through a variety of nongovernmental organizations, disability-specific or cross-disability youth councils or leadership programs that are oriented at least in part toward providing peer-support networks for youth with disabilities (although these are not necessarily oriented toward educational contexts, and would definitely not be focused on any specific field of study, let alone anything in STEM). Several grant-funded projects have been created to provide formal networking and peer-support opportunities to students in STEM; however, these have yet to be sustainable in an international or truly connective sense in the long term.

Thus you may need to create your own peer-support opportunities. If accommodation professionals and educators are aware of resources locally or online, this may help students create peer-support frameworks. Ultimately, building and subsequently maintaining a peer-support framework will be an exercise in networking (see chapter 16: Leveraging professional development and networking opportunities).

Formal and informal peer-support groups may be created locally, in your department or campus, according to the rules governing student groups at your school. Alternatively, a group of friends may become your peer-support network. Indeed, tapping into your existing social network is the best approach to building a peer-support network, depending on where you and your friends are located, and the strengths of these personal relationships.

Benefits of having a peer-support network

A peer-support network, no matter its form, can yield a number of benefits. First, peer-support groups, or a peer-support network, can provide a "safer space" for honest discussions about barriers, challenges, and potential solutions. A shift in conversational tone and perspective is created by peer-support networks, away from the "outsider perspective" of the student (particularly given the common social isolation of students with disabilities in STEM course and programs), toward a shared, more egalitarian way of talking and thinking. In a peer-support framework, a STEM student with a disability may find others who share common experiences. In a peer-support system, there is less of a need to use a"filter" for the "outside world." Finally, peer support has the potential to evolve into shared relationships between individuals with common experiences—friendships may develop, and the very nature of a student's peer-support framework may evolve.

Potential challenges in having or maintaining a peer-support network

Despite the many benefits of a peer-support framework, there are a few drawbacks to consider. First, peer-support frameworks may not be comfortable for everyone. The openness and trust required in a peer-support context may be difficult to accept initially. Some students may prefer interacting in smaller group settings or one-on-one in a close circle, instead of in a formal peer group, and therefore the choice of type of peer-support framework is often dependent on a student's comfort levels.

Additionally, some students may not be comfortable opening up in a large group setting, or may have a lack of desire for group interaction (as opposed to one-on-one). A student may also feel like the group has insufficient common ground with their own experiences to be an effective support, or may not trust the group as a whole, or some of its individual members. A peer-support group may have previously confused the boundaries between student and educator or professional, or between "peer" and "facilitator." As a result, the group may no longer be effective. For example, if the group is facilitated by a student services office or a guidance office at a university, college, or high school, it is possible the group's leaders or members may choose to emulate the professional behaviors and distance of the staff advisors, in potentially negative or damaging ways, rather than engaging as equals with students in the group.

Depending on the age of the group (how long it has been in existence) and member turnover, there may be divides between old and new members, and changes to the ideal egalitarian power dynamic as a consequence. The quest for similarity across different student experiences may also lead to competition. Alternatively, peer-support frameworks may become enabling spaces where students are not challenged to overcome their struggles. Cross-disability environments may have greater potential for disability-related conflict. As a consequence, it is important to maintain the principles of accessibility and universal design in the space defined by peer support. Finally, the student's motivation for engagement with the peer-support framework may also be a concern—is the student looking for something that the space cannot provide?

Operational issues: ensuring effective peer support

A peer-support framework in a group setting works best when there are a variety of levels of experience in the group, and when that experience is transformed into a viable knowledge base (what has worked in the past, what approaches have proved challenging, coping strategies, accommodation tactics, etc.). A peer-support group works best when it is able to offer and receive support in a variety of ways and is responsive to the needs of its members. Furthermore, peer support is more than the telling of and listening to stories—effective peer support needs to be able to offer analysis on a range of specific topics, including wellness and coping. Analysis in the context of peer support is as or more important than the stories of the group members themselves.

A peer-support group is more successful when it has experienced facilitators who are able to draw out the effective participation of all members and who are grounded in experience (and thus are less likely to be authoritarian in the context of the group). An experienced facilitator ensures the peer-support group has members who can help set the boundaries for the group. Group members with less facilitation experience should also be encouraged to share their ideas; those suggestions should then be evaluated by the group based on its needs and comfort levels.

Peer support versus peer mentorship

It is surprising to note that some people believe they can have a peer or a mentor but not both, which often results in missed opportunities for positive mentorship. Being exposed to the experiences of others, including your peers, can benefit you in the long run, even if you don't realize it at the time. By not restricting ourselves to the definition of a mentor as a "superior being" we can work toward a much more sophisticated understanding of the world and the value of the experiences of others. This viewpoint implies a "one-way" transfer of information during a mentorship relationship, when in reality things may be more nuanced (we return to a more wholesome discussion of mentorship in chapter 13: Faculty mentorship of students with disabilities in the sciences). Indeed, peer mentorship, can offer the opportunity for both parties to engage in the peer-facilitated learning process.

Importantly, peer support implies a two-way relationship or (in this context) a two-way mentorship process. Your peers are likely learning as much from you as you are from them. It is also important to remember that peer support and peer mentorship do not always involve positive experiences. A significant amount of literature supports the notion that we, as human beings, learn and process more from our failures, and those of our peers, than we do from our individual or collective successes. Thus an essential component to peer support and peer mentorship is the candid discussion of approaches, relationships, and trials that did not work out in our favor. (In the narratives to this book, an experiment in remote peer support, you have read some candid tales of challenge and failure, as a result of this philosophical approach.)

Peer support: the beginning of the conversation

Peer-support groups provide a middle space between support and formal advocacy— indeed, a number of disabled students' groups on university and college campuses arose out of peer-support settings. This evolution from peer support to advocacy— even the hybrid model, where a group continues to play both support and advocacy functions—provides a context wherein the student can be a part of "something bigger" and where the student may also gain comfort in approaching disability, accommodation, and disclosure, both personally and from a group advocacy perspective.

Through engagement in peer-support networks, a STEM student with a disability may gain a measure of "professional disability pride" in the context of their discipline, in that they become encouraged to engage more fully in the ACTion model presented in chapter 7, Student as ACTor—recognizing the importance of advocacy, communication, and trailblazing to student success in STEM.

Peer-support models and frameworks can therefore have a powerful impact on young scientists with disabilities. Several of these groups now have an online presence (e.g., PhDisabled in the United States and Chronically Academic in the European Union). Coupled with the "success stories" databases maintained by the American Association for the Advancement of Science and the National Educational Association of Disabled Students, the potential for growth and expansion of these networks—and for their eventual impact on research and collaboration—is significant. Peer-support systems for young scientists with disabilities are breeding grounds for collaboration.

Where do we find our peers?

Formal peer-support frameworks can be considered as an exercise in community development, while an evolution toward more informal peer support avenues can be considered as a manifestation of that nascent community.

The members of our peer group can originate from anywhere. We may meet them in hotels, bars, on buses, in restaurants, even in business settings or at parks. We may find them in our childhood, and have known them for years, or we may run into them in the first class of the current term. We often find these people when we are not looking for them or do not know what we need. You can begin as friends and then segue into a peer-to-peer support relationship based on common interests. There are many different kinds of peer support, and your peers can support you in many ways. The distinctions between the different types of peer support and different peer support roles are often nuanced, and can be unclear. Peer networks also allow you to mentor others, which in and of itself is a rewarding and even therapeutic experience. These networks can be a key tool in your success kit.

While we may think of peers as those closest to us in age, our peers can be of all ages and can be from anywhere in the world. For example, if someone with a rare disability cannot find anyone locally who shares their disability, a global network may be a good solution. Something as simple as a social media group dedicated to the experiences of a group of people who share a disability can provide the support and education a student needs.

By interacting with an online peer group as well as using more than one peer-support network (local as well as international), a student can take advantage of different opinions, mindsets, and even cultural biases. Peers are people whose opinions and ideas you trust and who you work with or share similarities with on one or many subjects. Using this definition, peer-support networks can be built based on hobbies, geographical location, or even employment status. Peers are people whose opinions and ideas you trust and who you work with or share similarities with on one or many subjects. Using this definition. peer-support networks can be built around hobbies, geographical location, or even employment status.

Virtual (long-distance) peer support

There are opportunities to build peer-support networks around every corner. It is not uncommon for people to build bonds at conventions or destination meetings and then to further foster the relationship after returning home. These virtual peer-support networks fostered from networking events can go a long way in supporting people dealing with social barriers or other hurdles. However, one of the most fundamental modus operandi of a network is to connect people to other people through mutual acquaintances. These connections can happen for any reason and at any time. Additionally, the outside perspective of someone who is not professionally or personally involved with your colleagues or coworkers provides the added benefit of not placing your professional relationships in jeopardy. By confiding in someone who is separate from your situation, you can expect a more honest perspective and the opportunity to speak more freely about your challenges.

Peer support outside disability

Peer support and mentorship go hand in hand and you are likely to learn more from your peers than you are from a teacher. Because our peers experience things at more or less the same time as we do, learning from their experiences and mistakes makes it much easier for us to navigate our own course. Peer support in the disability space is often, but not always, between people who have the same or similar disabilities. People who face the same or similar challenges often group together to share experience and solutions to problems. However, because of the nature of disability, it is important to have peer-support networks both inside and outside of the disability space.

While students with disabilities need friends who face similar challenges, it is equally important to have strong connections with nondisabled peers. Just as your disability does not define you, it does not define your relationships with people without disabilities. Without healthy relationships with others that do not share your disability, it is easy to create an "us versus them" mentality. Having peers from different backgrounds can give you a broad perspective on things and can also help keep you psychologically balanced.

Some of your peer-support networks (yes, you will have more than one) may be found in your community, but others may based online with members from around the world. While it is important to maintain networks with people close in age and location, it's more important to find others you can you connect with, even if they're not local.

Conclusion

When it comes to building and selecting peer networks, quality is notably more important than any single standalone factor such as geographic location, age group, or field of study. One of the greatest advantages of building your own peer network is that you have a greater degree of control over who you allow to be most influential over the course of your study, career, and life. This is notably different than some educational or employment settings where there is little you can do to select your teachers, classmates, or coworkers. These relationships can be national or international, and even transverse culture, field of study, age, etc.

As a person with a disability, your peer networks should not be limited to those in your field, those that share your disability, or those that reside near you. Differences in disability status, field of study, culture, and life experiences can be very valuable. Being a trailblazer is not easy, regardless of your field of study, but peer mentors can help you navigate as well as provide a more neutral sounding board than coworkers or fellow students in your immediate program.

Such relationships will prove valuable throughout your course of study and career. It is also important to remember the psychological benefits of being able to reach out to people outside of your immediate study or work environment. As people with disabilities, it is often assumed that we will need more help than we can give in return. Mentoring relationships allow for reciprocity, which is important in all relationships. It is not enough to simply feel as though you "fit in" within a community—you also need to know that you are making valuable contributions. Peer-support networks offer meaningful opportunities to fulfill this role.

Part IV

Student as Learner

Part IV

Student as Learner

Essential requirements and academic accommodations in the sciences

Chapter Outline

Introduction: what is an essential requirement?   119
Measurement of essential requirements   120
Importance of essential requirements   121
Essential requirements and the evolution of the sciences   121
Relationship between essential requirements and accommodation   122
Conversations around essential requirements and accommodations   123
Reasonable accommodation and undue hardship   123
The "gatekeeper function"   124
Mythbusting accommodations and essential requirements   125
    Myth: Accommodations are expensive   125
    Myth: Accommodations take away from understanding of course or program content   125
    Myth: Accommodations are synonymous with "having someone else do the work for you"   125
    Myth: Accommodations are unfeasible to implement for just one person   126
    Myth: Accommodations do not mirror the "real world"   126
    Myth: Accommodations will compromise safety   126
    Myth: Accommodations will interfere with timeliness   126
Essential requirements in STEM environments   127
Conclusion   128

"What does it mean to be a scientist?"

This wasn't a question I dwelled on much as a child. Having decided that being a scientist would be the best thing ever from a young age, I never troubled myself with that level of detail. What does it mean to be a scientist? It meant having a free license to think about stuff, to let my imagination go, to be creative, to ask really big, really important questions, and then to figure out their answers. Beyond that, who cared?

Everyone cared, as it turned out.

As I started to bend my educational path toward this dream of mine, I began to realize that not everyone was as enthusiastic as me about the prospect of me being a scientist – particularly when I started to think more about the experimental sciences when I started my postsecondary education. (I'd wanted to be an astronomer for a long time, but my interests evolved throughout high school and

Creating a Culture of Accessibility in the Sciences. DOI: http://dx.doi.org/10.1016/B978-0-12-804037-9.00010-3

*into my undergrad, and I ended up veering away from a field I was
passionately in love with, but highly theoretical in nature, to genetics
– another field that I was passionately in love with, but the farthest
thing from theoretical I could possibly find. Not that I didn't love the
theoretical sciences – in theory, I did. In practice, computer science
and I never got along all that well.) Educators began to ask hard
questions: "How, exactly, do you plan on doing this?" "If you need
help to do this, is it still your work?" "How do you intend to build a
career, if you need help to do this a certain way?"*

*All great questions – none of which I had previously thought about
until confronted with them and, in effect, was forced to do so. It was
then I realized that to address these questions, I really had to seek the
answer to the more fundamental question: "What does it mean to
be a scientist?" Put it another way – what does a scientist have to do
that is so essential to their job and role in society that if you couldn't
do it (or couldn't do it unaided), then you weren't a scientist?*

*One answer seemed obvious: If I couldn't think through a problem
on my own, I couldn't be a scientist. This was my childhood answer
revisited in some ways – after all, if I couldn't think about a problem
effectively, where would be the fun in being a scientist? Also, this
answer seemed to work for Stephen Hawking – perhaps the world's
most famous living scientist with a physical disability.*

*What about if I couldn't physically do an experiment? I'd chosen to
go into genetics – not the plant breeding that Gregor Mendel did at
the beginning of the field, mind you. Modern genetics involves direct
manipulation of DNA and the use of literally microscopic volumes of
liquid. For me, given my sight impairment, literally not being able to
see what I was doing was a significant and very real concern.*

*It took me some time to puzzle all this out, but it turned out that
this very question was at the root of the challenges my educators and
service providers were facing. Confront ten different lab scientists
with the question, "What is the minimum essential set of skills you
must have in order to be a scientist?" and you might get ten different
answers. Sure, there would be some overlap – but that overlap would
be mostly about critical thinking, troubleshooting and analytical
reasoning. What does a scientist physically have to do to be a
scientist? Never mind different answers in different disciplines – many
would give different answers even in the same discipline.*

*My doctoral research was in cancer genetics and involved a
mouse model of leukemia. I needed technical assistance in order to
perform the animal surgeries – but the goal of my research wasn't the
surgeries themselves, it was the analysis that needed to be performed
on the mouse tissues after surgery. I needed to be evaluated on that,
and I needed to demonstrate my competency to perform that analysis
– I did not, however, need to demonstrate my surgical skills in order
to graduate.*

This example illustrates the sort of thinking that admissions committees and the faculty who worked with me needed to go through. Once I got into my programs of study, I was in a position to think these issues through with them – after all, I knew my capabilities best, and could judge (with trial and error, if necessary) what I could or couldn't do on my own. A major challenge would arise if the admissions committee attempted this thought process – however subconsciously – on their own, without the benefit of understanding how my disability interacted with the learning or lab environment. They wouldn't have enough information to make an appropriate judgment – but weren't necessarily in a position to consult those who could help them to fill in those blanks.

So, what does it mean to be a scientist?

Everything I thought it meant as a child – with myriad practical conundrums to go alongside. Conundrums which require careful thought and which force us to confront our preconceptions about "doing science".

Introduction: what is an essential requirement?

"What does it mean to be a scientist?" The answer to that question is inextricably linked to the so-called "essential requirements" or "necessary competencies" in the context of scientific disciplines, which define the minimum set of skills and competencies a student must demonstrate in his or her program or field of study. In general, essential requirements can be thought of as learning objectives, but as you will learn in this chapter, there are some significant differences as well. As a student with a disability in the sciences, the issue of essential requirements is connected to the student's accommodations and whether these accommodations interfere with fulfilling the essential requirements of the course or program. These issues are also strongly linked to the challenge of meeting academic program requirements in the sciences for students with disabilities.

Thinking about academic requirements more creatively and inclusively is necessary in order to determine what aspects of a course or program content are "essential requirements" for completion of the curriculum, and which aspects can be accommodated. Educators must provide accommodation, up to the point of undue hardship, to enable students to meet these "essential requirements" (Ontario Human Rights Code, 2003, p. 62).

Rose (2009) notes that the essential requirements of a course or program "include (but are not necessarily limited to) the knowledge and skills that must be acquired or demonstrated in order for a student to successfully meet the learning objectives of that course or program." Oakley, Parsons, and Wideman (2012) describe two factors in identifying or defining the essential requirements of a program of study:

1. An essential requirement is a skill that must be necessarily demonstrated in order to meet the objectives of a course; and
2. An essential requirement is a skill that must be demonstrated in a prescribed manner.

An "essential requirement" is also a *term of art* in human rights legislation; i.e., it has a specific meaning beyond the commonly understood definition. An "essential requirement" in this sense is something that must be demonstrated in a particular fashion, with or without accommodation. In other words, it may be something that legitimately *cannot* be accommodated, because it is necessary that the person perform the task in a particular way or using particular methods or equipment, which could preclude a given accommodation. Such requirements are also referred to as "bona fide occupational requirements" (BFOR) in the context of the workplace; they are real, authentic, immutable, and nonnegotiable without compromising the inherent nature of a task. Bona fide occupational requirements are requirements that are shown to be:

1. Rationally connected to performance of a job;
2. Adopted in the good faith belief of their necessity; and,
3. demonstrably necessary, to the point of incurring undue hardship. (Alberta Human Rights Commission, 2009)

In the context of the academic (classroom or laboratory) setting, these could be understood as "bona fide academic requirements" (BFAR) (Roberts, 2013, p. 153). The connection between these two concepts is very important; it proves to be a conceptual challenge for educators and accommodation specialists in higher education, particularly in science, technology, engineering, and mathematics (STEM). "This accommodation won't work in an industry lab (or workplace setting), so why should we allow it here?" is a common, if inappropriate, refrain. How do we know that the accommodation won't work? In point of fact, we don't actually have that knowledge most of the time.

Measurement of essential requirements

There is a large difference between those requirements considered truly essential vs. those that may not be essential, but which are viewed as such, and this distinction needs to be understood and clarified. Educators and service providers need to understand and think through the requirements and whether each is critical to the understanding and practice of the course, program, or discipline. There is a danger of conflating the measurement of the requirement with the "actual requirement." For example, students with visual or physical disabilities may not necessarily need to "physically" carry out the experiments in the laboratory, but they do need the technical and scientific knowledge to instruct an assistant to carry out the precise measurements of the experiment. What is "essential" here is not the physical "doing" of the experiment, but the knowledge, skills, and direction the student requires to successfully see an experiment through (Sukhai, Mohler, & Smith, 2014). Educators, service providers, and administrators can often confuse the technical "doing" of a task in a particular way with its actual skill or competency. This is largely due to thought steeped in academic tradition that students must complete course requirements in a particular way. It becomes a problem when the method by which a task is accomplished is perceived as being more important than the quality of the final product.

Importance of essential requirements

The Council of Ontario Universities states that "learning outcomes are used to align individual courses with degree level expectations... (and) define what a student should know, and be able to do, after successful completion of an assignment, activity, class, course or program" (Roberts, 2013). In order to identify an essential course requirement, "several questions can be applied to the requirement that help refine the rationale for its inclusion...." These include:

- What is being tested?
- What is the nature of the task?
- Does it have to be done in only one way?
- If so, why? (Roberts, 2013, p. 52)

When looking at whether tasks can be completed in an alternative way, it is necessary to examine whether modifying the way a specific task is completed will compromise the student's ability to achieve the objective of the task (Roberts, 2013). For example, does a student who is blind need to physically hold the pipette in a chemistry lab, or can a lab assistant be employed to carry out visual tasks in the lab, as directed by the student? "If the objective of the task can be achieved with the use of an accommodation, the method of execution is not an essential requirement for the task" (Roberts, 2013, p. 52). While requirements for testing of course material, labwork, and assignments have been pre-determined, it is necessary to explore whether tasks have traditionally been completed using a specific method to achieve a course objective, or whether alternative ways of completing tasks can be considered to achieve the same objective.

Developing creative examination techniques of the essential requirements for the course or program cannot only serve to encourage evaluation of traditional testing techniques, but can also introduce new methods of testing to instructors. For example, Miner et al. (2001) noted that not all testing needs to be completed in written form. They explained that "oral tests, presentations, and group work/projects are among a rich assortment of alternative methods to evaluate course understanding" (p. 45).

Essential requirements and the evolution of the sciences

How we teach and our mindset surrounding why we require tasks to be done in a particular fashion impacts teaching and learning outcomes for all students, but especially students with disabilities. The pace of knowledge growth in the sciences today far outweighs advancements and changes in teaching methods and practices. Thus the definition of essential requirements is unclear for really new (e.g., bioinformatics and the data sciences) and/or rapidly evolving disciplines (such as genetics). Thus there is a lack of consistency of thought and application around essential requirements among professionals in emerging fields.

The technical differences between fields also translate into significant variances in essential requirements definitions. Consistent conversations around essential requirements are more likely in professional fields than in basic science disciplines (such as biology and chemistry). It can be difficult to restructure or think creatively about accommodations because of traditional approaches to instruction. Changing methods of course and program evaluation can be seen by educators as changing the requirements for a given discipline or program, rather than as simply exploring alternative means of completing course requirements (Sukhai, Mohler, & Smith, 2014).

Relationship between essential requirements and accommodation

Accommodation "is a means of preventing and removing barriers that impede students with disabilities from participating fully in the educational environment in a way that is responsive to their own unique circumstances" (Ontario Human Rights Commission (OHRC), 2004, p. 7). Similarly, the Alberta Human Rights Commission (AHRC) (2010) states, "Accommodation does *not*:

1. Require that postsecondary institutions lower academic or nonacademic standards to accommodate students with disabilities.
2. Relieve the student of the responsibility to develop the essential skills and competencies expected of all students."

Appropriate accommodations enable students to meet the essential requirements of a program successfully, "with no alteration in standards or outcomes, although the manner in which the student demonstrates mastery, knowledge and skills may be altered" (OHRC, 2004, p. 24). This would give all students "equal opportunities to enjoy the same level of benefits and privileges and meet the requirements ... without the risk of compromising academic integrity" (OHRC, 2004).

Evaluation accommodations are determined on an individual basis. Accommodations do not provide an advantage for the student; rather, they help the student compensate for the effects of his or her disability. The critical factor in providing reasonable evaluation accommodations is determining the essential evaluation components of a course and the extent to which accommodations are appropriate for a student with a disability. To determine this, the following questions may be asked:

1. What methods of assessing performance are absolutely necessary, and why; and,
2. What alternative methods of evaluation could be considered to determine the student's appropriate level of knowledge or skills?

The disability services provider evaluates each student's needs and make recommendations regarding test/examination accommodations. For example, students may receive extra time if it takes them longer to read, write, or process the test material. An oral exam may be given instead of a written exam if the student has difficulty writing or seeing.

Other frequently used examination accommodations include the use of technology or special software; students with visual or physical disabilities or severe learning disabilities may need a reader or scribe to assist them in completing the test questions. It is important to note that the reader or scribe does not assist the student other than by reading the questions aloud or copying down exactly what the student dictates to them (Oakley et al., 2012).

There are times when students should not be accommodated, such as when altering the method of performance changes the fundamental nature of the task. Both the immutable elements of an academic task or requirement and the functional impairments of the student must be considered in concert to find the appropriate accommodation, or to justify denial of an accommodation (Roberts, 2012).

Conversations around essential requirements and accommodations

Academic performance of the student, provision of the appropriate academic accommodations, and essential requirements are often very closely linked—so closely linked, in fact, that educators and service providers find it challenging to have conversations with students with disabilities about their grades. Indeed, anecdotal evidence suggests that a certain level of avoidance is practiced around these types of conversations, which ultimately does no one—student, educator, or service provider—any use, as these cases end up being presented to university academic appeals boards. Involvement of the academic judiciary process can escalate tensions and leave students and educators with mindsets unconducive to future participation in STEM for people with disabilities. Such challenges can be avoided upfront, with some careful thought and willingness to engage in dialog among the student, educator, and service provider. The following scenario offers a case study on how to accomplish this.

Case Study: A student with a disability is obtaining a failing grade in a physics laboratory course that is part of their program requirements.

1. If the student is failing because the disability makes it impossible to meet certain course requirements, examine the importance of the problematic requirements in your course.
2. If the problematic requirement is not essential, you may want to adjust your grading scheme so that the student can show mastery of course material in equivalent ways. (Note that this is *NOT* the same as simply waiving a requirement—it is *replacing* it with one of equivalent importance and level of difficulty.)
3. If it is essential that the student be able to complete all requirements without accommodation, let the student know early in the semester that the course requirements will not be modified.
4. If modification of the course requirements is not possible, let the grade stand, after discussion with the student and service provider.
5. If the student fails for the "usual" reasons, allow the grade to stand. Do not pass students with disabilities just because they tried hard. This is unfair to all of your students.
6. You may or may not want to speak to the student about the failure in this latter situation.

Reasonable accommodation and undue hardship

The duty to accommodate requires that accommodation be provided in a manner that "most respects the dignity of the person, if to do so does not create undue hardship" (OHRC, 2000, p. 10). The OHRC's Guidelines on Accessible Education (2004) state that "every student with a disability is entitled to accommodation up to the point of undue hardship." The determination of essential requirements is independent of the determination of undue hardship. Only three elements may be considered in assessing whether an accommodation would cause undue hardship: cost; outside sources of funding, if any; and, health and safety requirements, if any (OHRC, 2004, p. 28). The evidence "required to prove undue hardship must be objective, real, direct and, in the case of cost, quantifiable" (OHRC, 2000, p. 24). The institution cannot argue undue hardship based on business inconvenience, employee morale, third-party preference, or collective agreements or contracts (OHRC, 2000, p. 22).

The "gatekeeper function"

Too often, students are discouraged from going into STEM because faculty and instructors are not aware of alternative ways to work in the lab environment, or of alternative means of "doing." Many educators think having a student with a disability in the laboratory could pose risks to safety, or that the student would not be able to physically handle the demands of the labwork. There is little in the way of role models for students with disabilities in STEM, and this presents challenges when students need to seek out information that would better enable them to advocate for their own learning.

Students may take on the role of "mentor" as they can play a large role in educating faculty, department heads, and course directors about the interaction between their disability, accommodations, and the course content. This can happen by the student giving presentations to faculty and staff and engaging in open dialogue with faculty and course instructors. By doing so, the student provides some education about disability management and their accommodation needs to their faculty and department.

Students may educate faculty and administrators in their field on the impact of a disability on a student's participation in the field. This may happen through the training of faculty and lab staff around disability and accommodation, as well as essential requirements and creative and flexible solutions to accommodating students with disabilities. It is also critical to foster open and collaborative communication among faculty, administrators, service providers, and students around developing creative and flexible solutions to accommodations and essential requirements. It is easier to teach technical staff about accommodation than it is to teach disability services and accommodation staff the technical issues inherent in the sciences, given the sheer weight of technical knowledge required across a multiplicity of disciplines.

With the right mindset, solutions to essential requirements problems that are practical and enable the student to meet the appropriate requirements can be found. This requires collaboration and critical approaches to thinking through the problem. There are not many students with disabilities found in STEM (see chapter 1: The landscape for students with disabilities in the sciences), so students may end up as pioneers, trailblazers, or role models without expecting it, or knowing how to cope with this newfound responsibility (see chapter 7: Student as ACTor—recognizing the importance of advocacy, communication, and trailblazing to student success in STEM).

Students may need guidance on how to advocate for their educational needs and to approach faculty (which is crucial for sorting through essential requirements and accommodations), and some may need training on how to address these unique issues. Who teaches students this skill depends upon the school, department, and the student's needs. The student may have to crowd-source people through their networks or through available resources and existing role models. Advocating for accommodations can provide students with valuable opportunities to gain skills, effect positive change, and expand their academic and professional networks.

Mythbusting accommodations and essential requirements

In our experience, a number of myths have grown up around accommodations and essential requirements—particularly in the sciences, where there are few systemic opportunities to counter these myths. Here, we provide some STEM-specific counterexamples and underlying principles to aid in countering these common myths that serve as barriers to participation for students with disabilities.

Myth: Accommodations are expensive

Counterexample: Handheld magnifier for a visually impaired student to assist with reading labels on chemical bottles, as opposed to printing large-print labels on an ongoing basis.

Principle: 90% of accommodations cost less than $500 to obtain and implement.

Myth: Accommodations take away from understanding of course or program content

Counterexample: Alternative evaluation methods (e.g., oral exams) provide means for the student to communicate course and program content in a way that facilitates their independence.

Principle: Accommodations "level the playing field" for students and allow them access to tools that facilitate inclusion in a course and program.

Myth: Accommodations are synonymous with "having someone else do the work for you"

Counterexample: Experiments are designed and directed by the student with a disability who then also analyzes and interprets the results.

Principle: Design and interpretation of a scientific experiment are core skillsets, and these can be accomplished by a student without technical assistance. Technical assistance occurs with specific direction by the student, meaning that the student must have intimate and detailed knowledge of the steps of the procedure as well as how to troubleshoot any errors. While many lab-based courses require the student to physically carry out the experiment on his or her own, it is actually more time-intensive and challenging to direct another individual to do the precise and accurate steps required in labwork. This method more effectively demonstrates the student's understanding of the methods, techniques, and the objectives of the experiment.

Myth: Accommodations are unfeasible to implement for just one person

Counterexample: Accommodations are determined, in large part, on a case-by-case basis, which therefore contributes to the overall individual nature of accommodations. Individualized accommodation plans not only benefit the student for whom they are intended, but can serve to inform educators and course instructors (and set precedents for future students with disabilities in the lab setting).

Principle: Due to the nature of many graduate, professional, and field-based programs, individual accommodations based on the student's specific disability and the program of study are more appropriate than a standard accommodation checklist. For example, students in graduate education may require accommodations around their research work that need to be creative and flexible given the nature of the work.

Myth: Accommodations do not mirror the "real world"

Counterexample: Students with disabilities are aware of their abilities, and how these abilities may play out in their professional role after education. Students will often self-select into fields of employment that fit with their skillsets and abilities, thus potentially minimizing their accommodation needs.

Principle: It is critical for faculty and administrators to avoid "gatekeeping" students from entering specific fields of study because of perceived barriers surrounding the student's ability to complete the degree requirements.

Myth: Accommodations will compromise safety

Counterexample: It is often assumed that a person with a disability will not be safe in the lab. However, it is the lack of appropriate integration between disability, accommodation, and the lab that may lead to safety challenges. Appropriately implemented accommodations—particularly those designed upfront and in a proactive manner—are designed with lab safety guidelines in mind, and as a best practice, ought to be done in full consultation with the appropriate health and safety committee. Physical modifications may be made to the built environment, which take safety into account, and such modifications work to prevent any contact with hazardous materials and substances. For example, students with visual or motor coordination disabilities may be exempted from working with radioactive materials in the lab.

Principle: Appropriate accommodation goes hand-in-hand with lab safety, and best practices will take both the accommodation need and the relevant safety issues into consideration.

Myth: Accommodations will interfere with timeliness

Counterexample: Many scientific procedures need to be done in specific timeframes; otherwise, the experiment may not work as designed or intended. Students with disabilities—particularly those with motor coordination and focus difficulties—may take longer to complete these tasks without accommodation, thus interfering with the success and completion of the experiments. Here, accommodations to address these issues actually help with, as opposed to hinder, the student's timeliness in a lab setting.

Principle: It is important to carefully think through each step of an experiment to determine where a lack of accommodation may compromise the student's ability to fulfill the requirements of the task.

Essential requirements in STEM environments

At the graduate level, the identification of essential requirements has begun mainly in those professional graduate programs with explicitly defined requisite skills and abilities or competencies, licensure requirements, and postgraduate-associated regulatory bodies (Barker & Stier, 2013). Furthermore, the determination of core competencies in research-stream programs has been idiosyncratic rather than systematic and dependent on the faculty member or department. One of the significant challenges in determining essential requirements in the graduate environment is that most graduate programs have not formally or sufficiently articulated their program goals, core competencies, and learning outcomes in outcomes-based language. The accommodation process then has to include analysis of both program requirements and accommodations.

In the context of graduate professional education, program requirements are thought of as those skills required for qualification in the discipline. These may include:

1. "General" necessary requirements, which are applicable across multiple disciplines (e.g., the ability to perform critical analyses, form testable hypotheses, use professional judgment, establish rapport with clients/patients);
2. Discipline-specific requirements (e.g., technical writing, specific forms of data interpretation, standardized test administration, clinical interviewing); and
3. Technically necessary requirements (e.g., use of specific methodologies, materials, tests and/or instrumentation).

Of note, various licensing and regulatory bodies define the core competencies in a range of professions, particularly those associated with end-of-program licensing requirements such as medicine, nursing, pharmacy, occupational and physical therapy, law, and social work. Professional graduate programs in these fields work within this licensing context, teaching to ensure successful completion of licensure examinations. Many of these programs are also developing explicit expectations of requisite skills and abilities within their respective fields (University of Saskatchewan, n.d.; Council of Ontario Faculties of Medicine, 2003; McGill University, 2013; University of British Columbia, 2013; Nursing Association of New Brunswick, 2014).

However, the majority of research-stream master's and doctoral degree programs do not have analogous qualification requirements; indeed, each field is likely to have numerous affiliated professional societies (regional, national, international), and there is usually no consensus on definitions of core competencies for research-stream graduate programs. The unique nature of particular degree programs and the various ways in which individual disabilities intersect with the requirements result in a plethora of potential accommodation needs, with no two the same. This dynamic interaction between program requirements and individual disabilities makes it difficult, if not impossible, to determine universally applicable requirements.

Interestingly, a similar challenge exists at the postdoctoral level; there, some best practices have evolved through the work of organizations such as the National Postdoctoral Association (NPA) and the Federation of American Societies of Experimental Biology (FASEB). The NPA published a list of core competencies defined as a series of general necessary requirements that are recommended as learning outcomes from postdoctoral training, while FASEB utilized this framework to develop material for research trainees to use in evaluating their development of these competencies. In parallel to the NPA, Vitae in the United Kingdom evolved a Researcher Development Framework that serves a similar purpose as the NPA's Core Competencies (https://www.vitae.ac.uk/researchers-professional-development/about-the-vitae-researcher-development-framework). These materials are intended to be cross-disciplinary and are adaptable to the graduate environment. There is a similar effort by the Tri-Council granting agencies (the Social Sciences and Humanities Research Council, the Natural Sciences and Engineering Research Council, and the Canadian Institutes of Health Research) in Canada to evolve an analogous research training competency framework. However, to the authors' knowledge, this effort never moved beyond the drafting stage and no documents have been published.

Given the wide variability of graduate programs, both research-based and professional, it is therefore critical to examine academic requirements more systematically and creatively to determine essential requirements for completion of the curriculum and performance in the field, which aspects can or cannot be accommodated, and why.

Conclusion

A critical issue for students in STEM disciplines is the interface between the student's disability, program, and required accommodation. It is necessary to separate the evaluation method from the actual competency being evaluated. Faculty are responsible for identifying the essential requirements in their respective courses, and it is therefore necessary for students to continually engage in open and frequent dialogue with their faculty around their accommodations and the requirements of the course. Through creative problem-solving and consistent communications, barriers to accessing STEM courses and labs can be ameliorated. It is important for faculty and departments to be flexible and explore alternate means of completing course tasks. As such, it is necessary for faculty to work with both the student and the disability service provider to develop appropriate accommodations to meet essential requirements.

Appropriate accommodations enable students to meet the essential requirements of their course or program successfully, with no alteration in standards or outcomes. In order to determine which aspects of the course or program are "essential," it is necessary to think more critically and inclusively about the academic requirements in STEM fields.

Universal design for learning

Chapter Outline

Introduction   130
Barriers to learning   131
Principles of universal design for learning   132
Applying UDL to the learning environment   133
The role of technology in implementing UDL   134
UDL and the need for accommodation   135
Conclusion   136

In the sciences, we're very accustomed to extremely didactic presentations. Particularly in graduate school, we don't attend lectures as much as we sit in on seminars – indeed, as instructors, many of us don't give lectures as much as we give seminars. A one-way download of information from us as the instructor or presenter, to the students and other trainees. To some degree, this is understandable – it's how we present information at conferences, and how we're taught to tell stories in the sciences. Perversely, once we get out of the undergraduate teaching labs, it's also how we're taught to learn.

Over more than twenty years in science, as a student, as teaching staff, and as a researcher, I've borne witness to some impressively egregious teaching styles. One faculty member would literally read his tremendously detailed handouts out to the class, word for word without inflection. Others would be at the exact opposite end of the spectrum, flashing incredibly dense and totally unreadable slides (even for the fully sighted students!) on the screen for a maximum of 15 seconds at a time, anticipating we'd absorb all the details by some miracle of knowledge transference. I've seen faculty members shy away from interaction during class, because it interrupted the flow of their presentation. I've seen other faculty refuse to consider alternative approaches to teaching, because they'd been doing it a certain way for their whole career.

None of these individuals employed universal design for learning approaches. Now, in their defense, UDL was, during my undergraduate education and my Master's, still in its infancy, and not very well known.

The only problem for me? I can't teach the way I was taught. Maybe it has something to do with having a disability myself, but I felt the urge in developing my teaching style to be more universal. I thrive on student engagement and interaction. The Socratic method is very much my thing. Problem-based learning. Case studies. Working with the students to adapt to their learning styles. Knowledge application exercises. Debates. Letting the students' natural curiosity and investigativeness flower and bear fruit in my classes. Teaching without the seminar based approach.

Taken to an extreme, that's probably not very UDL either – not all the time, at any rate. I don't have detailed lesson plans, as an instructor. I have learning objectives, and about a dozen ways I'd be happy if they got achieved. I want students to come out of my classes with knowledge sets, and I tell them that up front, but I don't much care if how we get from here to there doesn't follow my original plan.

I have slides, but 19 times out of 20, they won't get used in the order they were prepared, and they might even be used for a different purpose than originally designed if the students are on to something and I think it better to run with that. Some of that has to do with my own disability and my own learning style biases, because I had to evolve a teaching method that works for me, in the context of those things.

Students in the sciences either love or hate the UDL approach – I've taught classes where 80% of the class gets what I'm trying to do and engages with it viscerally and with fervor. And I've taught classes where 80% of the students don't go for UDL or the Socratic method at all – they want a seminar, and frankly think their time is being wasted otherwise. Graduate students, especially, are prone to this perspective – that anything other than a straight up didactic approach is for undergrads.

We teach what we learn.

We also teach HOW we learn. There's a lot of cultural sea change required for full adoption of UDL in the sciences, and it will take a while. There's so much benefit, though – not just to students with disabilities, but to all students and to the instructors as well.

Introduction

Universal design (UD) is a principle that is used in many fields, from engineering to art. UD ensures that what has been produced can be used by the widest group of people, including those who may have a disability. Universal design for learning (UDL) continues this practice into the classroom. There are three main components of UDL:

- Faculty represents information and concepts in multiple ways (and in a variety of formats).
- Students are given multiple different ways to express their comprehension and mastery of a topic.
- Students engage with new ideas and information in multiple ways.

Learning materials, methods, assignments, and evaluations are created to ensure that all students have the opportunity to participate fully. UDL is not a method of accommodating students with disabilities—it goes far beyond that. It incorporates a variety of methods that enable all students to learn and express their learning using an array of techniques (Burgstahler, 2008). One size does not fit all and UDL, through an instructor's utilization of multiple approaches to presenting the material, as well as the student's ability to express his or her learning in multiple ways, encourages flexibility in delivery and design of learning so that all students benefit.

UDL may seem challenging, particularly when dealing with large university classes. However, technology provides many opportunities for incorporating UDL principles into teaching and learning (Burgstahler, 2008). The faculty member does not have to create these materials, but instead can offer links to materials provided online, in textbook e-packs, or in learning object repositories, to name a few. In this way, the concept is explained or demonstrated in different ways, thus providing opportunities for more students to understand the concept (Center for Applied Special Technology 2011A; Center for Applied Special Technology 2011B; Centre for Applied Special Technology, 2011C).

Using UDL principles, students are given a choice regarding assignments (we know what you are thinking—this will increase the grading required!). The grading has to be done in any case, but offering students a choice allows them to choose the method that best demonstrates their comprehension of the concepts. That being said, if the purpose of the assignment is to learn how to write a report or develop a working prototype, other methods for completing the assignment, such as creating a video, may not be appropriate.

UDL evaluations work best when students are to demonstrate their knowledge of concepts (such as their understanding of particular issues or problems), or their ability to analyze particular issues. Providing alternatives to completing the assignment enables students to choose and determine how they can best demonstrate their competency to the instructor. This may mean writing a paper, creating a video, designing a poster, or making a model. As the instructor you are to determine how well they understand the concept, but the rubric or criteria used to mark each assignment can be the same.

Research has determined that active learning can increase a student's understanding of the concepts. When students are involved or engaged with the content—when they touch, react to it, analyze it, discuss it—it can provide deep learning. The more contact they have with the content, the deeper their understanding becomes.

Barriers to learning

Barriers to learning for students with disabilities can include attitudinal, organizational, or practical (Higbee, 2008). In higher education, particularly as classes get larger, the learning experience is based on transferring knowledge from the instructor to the student. The emphasis on print materials (even most PowerPoint presentations are predominantly print), and the dominance of one-way communication in the lecture hall, can make learning more difficult for many students, particularly for those who have a learning disability or other form of disability. It also does not promote deep learning, which can occur through learning activities that are collaborative, engaging, and designed to enable students to construct meaning from the experience.

Consider alternate methods of evaluation when creating assignments and tests; e.g., providing multiple choice tests may benefit a small portion of the class, and simultaneously disadvantage other learners (Higbee, 2008).

Other barriers to learning can include time and money. Does your course provide flexibility for students who have to work in order to go to school?

There are also cultural barriers to learning; language differences; learning styles (e.g., a kinesthetic or hands-on learner may have difficulty in lecture-based courses); attitudinal barriers, which can include confidence or motivational issues; and, of course, disability-related issues. For example, a student with a particular learning disability may struggle in a text-based online course because the student may need more frequent access to an instructor and peers for clarification of materials. These are just a few of the many barriers our students experience in postsecondary education. Incorporating UDL principles can go a long way to assist in the removal of these barriers and improve learning for the majority of the student body.

Barriers to learning vary widely (Higbee, 2008). Even if diagnosed with the same disability, the impact on students may be quite different. While most students with disabilities work closely with their disabilities office and advisor on skills and strategies, the faculty member should also discuss how best to meet the needs of students with disabilities. It is inappropriate to ask specifically about the diagnosis of any student, but conversations around assisting him or her are encouraged. Faculty and course instructors should also feel free to contact the disability services office regarding students with disabilities (Higbee, 2008).

Principles of universal design for learning

There are three overarching principles that guide the implementation of UDL:

- Multiple methods of representation that give learners a variety of ways to acquire information and build knowledge.
- Multiple means of student action and expression that provide learners alternatives for demonstrating what they have learned (Rose, Harbour, Johnston, Daley, & Abarbanell, 2006).
- Multiple modes of student engagement that tap into learners' interests, challenge them appropriately, and motivate them to learn.

UDL began with the universal design (UD) movement of the 1990s. The term "universal design" was coined by architect and designer Ron Mace at the Center for Universal Design at North Carolina State University (Burgstahler, 2008; Center for Applied Special Technology, 2009b).

Mace and his colleagues defined UD as "the design of products and environments to be usable by all people, to the greatest extent possible, without the need for adaptation or specialized design." Three critical insights that emerged from this work have come to define UD:

Most retrofitting and "adaptation" could have been avoided if designers had planned for accessibility from the beginning. Mace suggested a design ideal in which the needs of a diverse audience should be anticipated. Thus a chief characteristic of UD is that it "proactively builds in features to accommodate the range of human diversity" (McGuire, Scott, & Shaw, 2006, p. 173).

Modifications to the built environment—automatic door openers, curb cuts, entry ramps, universal-height drinking fountains, and others—are beneficial to many people, not just those with disabilities. Indeed, people today routinely use door openers to enter a building when their hands are full, just as skateboarders use curb cuts and children visiting the hospital can drink water from a fountain without assistance. Similarly, commuters in noisy airports and students in quiet libraries rely equally on TV closed captioning. Each of these conveniences was originally conceived as a disability accommodation.

Disabilities have less to do with individual deficits—what some people can't do that others can—and more to do with environmental barriers that obstruct people's ability to function effectively and participate fully in society.

In recent years, the UD philosophy has found fertile ground in the field of education. Elementary school teachers and university professors alike have adopted UD "as a conceptual and philosophical foundation on which to build a model of teaching and learning that is inclusive, equitable, and guides the creation of accessible course materials" (Schelly, Davies, & Spooner, 2011, p. 18).

If the goal of UD is the removal of barriers from the physical environment, the goal of UDL is the elimination of barriers from the learning environment.

The obstacles faced by students with disabilities (e.g., study materials that are not in electronic formats, uncaptioned video, PDF files that do not contain any real text and therefore cannot be searched or read aloud by text-to-speech software, etc.) are often the same obstacles encountered by students with different learning styles or whose native language is not English (Rose et al., 2006). UDL holds the promise of benefitting many students—hence the "universal" in UDL.

No single way of presenting information, responding to information, or engaging students will work for all students. Alternatives reduce barriers to learning for students with disabilities while enhancing learning opportunities for everyone.

UDL does not advocate any single teaching practice; rather, it combines today's best approaches for engaging students and challenging them to think critically. It helps instructors meet the learning needs of a diverse student body through a combination of instructional modalities, formats, and technologies. UDL is simply teaching in a manner that respects the needs of a diverse group of learners.

Applying UDL to the learning environment

When applying UDL to the classroom environment, it is important to keep the learning styles as well as the potential accommodation needs of students in mind. In terms of providing information, it is easiest to reach all students by varying the method of presentation and not solely lecturing. If possible, include videos, group work, role-playing situations, and class discussion. To eliminate the need for some accommodations, also consider providing copies of class notes, PowerPoint slides, and/or outlines (if available). This can reduce the need for note-taking accommodations in your class, while providing the benefit of these accommodations to each student. Providing these types of notes enables all students to focus more on the class and class participation as opposed to the mechanical task of note taking. It also helps students to better understand key points by potentially providing a different perspective on what is and is not important.

The advancement of computer technology has led to many class materials being created and distributed in digital form, making it easier for students to use. Having digital text is often the first step taken to implement UDL in a class. Digital text is a UDL consideration because it can easily be turned into Braille or enlarged in size for students with visual impairments. For students with physical disabilities who cannot physically manipulate the book in order to turn the pages, people with learning disabilities, English as a second-language learners, or simply people that consider themselves auditory learners, digital text combined with some type of text reader can read text aloud. By providing class materials such as text books, class notes, extra readings, etc., in digital form, the instructor allows students to access that text in whatever format works best for them, a true example of UDL. However, simply making text available in an electronic form does not guarantee accessibility. When a PDF is inaccessible (i.e., it cannot be read aloud by a screenreader), the text is often shown as an image. In fully accessible PDF files, each chunk of text is identified as text and images are given proper alternative text descriptions. Instructors need to be aware of and follow the steps to creating accessible PDF and Word documents before posting them to a class website to make the most of UDL.

The role of technology in implementing UDL

Technology plays a key role in implementing UDL in a class, and there are several simple technologies that instructors can use. For example, the University of Iowa is using a Livescribe Smart Pen (www.livescribe.com) to take and provide class notes in multiple formats. The Smart Pen is a pen that contains a recording device, which when used with its accompanying notebook, links written notes to what was recorded at the time the note was written. When reviewing notes, the user simply taps a written note with the pen and listens to the attached recording. Both audio and written notes can be transferred to a computer where they can be shared and saved. At the University of Iowa, several instructors are implementing the use of this technology to supply class notes in multiple formats to all students. Instructors assign a different student to take notes each class period. At the end of the class period, they collect the pen, extract the notes, and share them via the class website. With these notes on the website every student not only has access to written class notes, but can also access them in an audio format. Response to this tool has been positive, and in classes where it has been implemented accommodation requests for notes and notetakers have reduced. The Smart Pen is only one example of technology that could be used to help implement UDL principles in the classroom. Providing access to several other tools that are often considered assistive technologies such as free text readers can help with UDL implementation as well.

UDL and the need for accommodation

McGuire et al. (2006) note that "architects and designers implementing UD do not make claims of creating totally inclusive products and environments. They speak of designing products that are accessible to the greatest number of users" (p. 171). Similarly, while UDL is meant to cover a range of abilities "it is important to note that certain students will continue to require specific accommodations according to their individual needs above and beyond what UDL can achieve" (Dawson, 2004). In reality, faculty who embrace the practice of UDL or inclusive teaching practices allow "students with disabilities (to) have increased access to course participation with fewer special accommodations" (Shaw, 2011). In short, UDL may reduce, but not fully eliminate, the need for accommodations (Langley-Turnbaugh, Murphy, & Levine, 2004; OHRC, 2003; Shaw, 2011).

Chickering and Gamson's (1987) *Seven Principles of Good Practice in Undergraduate Education* has been widely acclaimed, practiced, and adapted for use in higher education. The seven principles for active learning complement inclusive teaching methods, in particular the principles of universal instructional design, in considering the needs of diverse learners (Brock University, 2012; Johnston & Doyle, 2011; University of Guelph, 2006). UDL is therefore "the philosophical foundation for inclusive teaching" (Moon, Utschig, Todd, & Bozzorg, 2011, p. 332). Unfortunately, UDL is not widely implemented in most postsecondary institutions. However, the OHRC (2004) indicates that in order to ensure that students with disabilities have equal access to education, academic facilities, programs, policies, and services must be structured and designed for inclusiveness. To avoid creating barriers, education providers have an obligation to be aware of the differences between individual students' learning needs, as well as the differences that characterize the diverse learning needs of students when making design choices.

Through accommodation and inclusive teaching practices, we know that "when students with disabilities are supported and made equal participants in our courses, they enhance the quality of the classroom experience for [the instructors], for themselves, and for their peers" (Johnston & Doyle, 2011, p. 53). In summary, UDL is a "positive approach (and) is more effective because it is accessible and inclusive from the start. Barrier prevention is much more preferable to barrier removal ... ." (OHRC, 2004, p. 10).

Sheryl Burgstahler (2012), Director of the DO-IT program at the University of Washington, provides a thorough review of the different approaches to UD in a laboratory setting in her paper "Making Science Labs Accessible to Students with Disabilities." She reports that students face challenges in accessing labs, and barriers in a lab setting can result in a lack of knowledge and participation. The two main approaches for making labs accessible relate to providing specific accommodations (alternative formats, adapted equipment, modified lab spaces for wheelchair access) and UD. "Accessible labs will also permit smoother teaching of concepts, as staff/ teaching assistants can teach to a common audience and not have to target specific subgroups" (Burgstahler, 2012).

Making accommodations, as required by individual students who have disabilities in science programs, is reactive. UD is a proactive approach to addressing most barriers in a lab setting (Hilliard et al., 2011). When considering UD in the lab, which will benefit all science students, different types of disabilities must be considered for accommodation, including blindness/low vision, deafness/hearing loss, mobility disabilities (in particular those who use wheelchairs or scooters), learning and attention difficulties, and health and mental health disabilities.

"Universally designed, accessible laboratories promote the inclusion of students with disabilities without adaptations. Many changes can be made to the physical environment, which are unnoticeable to many. Likewise, adaptation such as an adjustable height lab bench accommodates everyone and can have many benefits for persons without disabilities. They are allowed to stand or sit based on the type of work that has been done and can also alleviate repetitive stress" (Johnston & Doyle, 2011).

UD, as applied in classroom, requires educators to teach their courses and design curricula for diverse learners, including those with disabilities. An example of an academic accommodation that benefits persons with disabilities, but also supports the learning of others in the class, is the captioning of video/film presentations in academic coursework. Captioning benefits deaf students, but it is also a complimentary learning tool for other students who are visual and auditory learners or English as second-language learners.

Course curricula, delivery methods, and evaluation methodologies should be designed inclusively from the outset. This may require the creative use of technology, such as putting materials online or selecting software that is compatible with screen-readers. When online or other electronic content is offered, accessibility issues should be addressed upfront, in the development stage. Other examples include considering accessibility during the design of science laboratories and procurement of laboratory equipment. Indeed, many learning needs of students with disabilities can be accommodated with the creative adaptation of mainstream and off-the-shelf technology or equipment.

Conclusion

There are two broad ways of addressing accommodations for students with disabilities. One approach suggests that accommodations are individual in nature (e.g., the student has a disability that interferes with his or her ability to access the content of the course, to express knowledge, or to engage optimally in it). Such a view fosters solutions that address weaknesses in individuals. In contrast, issues can be considered "environmental" problems in the design of the learning environment. For example, overreliance on printed text for presenting content and evaluating students clearly, and differentially, raises barriers to achievement for some students while privileging others. Such an environmental view fosters solutions that address the limitations of the learning environment rather than the limitations of the student, while making the student less of a problem, and more a part of diversity within the course (Rose et al.,

2006). The advantage of such universal solutions is that, as with such approaches in built environments, they are likely to be useful for many individuals; i.e., built once, applied many times.

It is worth noting that the two approaches highlighted above are not mutually exclusive—even within the context of a universally designed learning environment, some level of personalized accommodation may still be required. Accommodation does not preclude the need for and benefits of UDL, nor does UDL exempt us from our responsibility to provide accommodation where and when it is necessary.

We believe that both approaches are important from a pedagogical standpoint. Moreover, in their intersection, we will find solutions that are not only more economical, but also more equitable. These two approaches reflect the fact that disabilities always reflect mismatches between the environment and the individual. Some universities place too much emphasis on disabilities within students, and not enough on disabilities in the learning environment. Accommodations and access issues are largely addressed on an individual basis, rather than on the level of courses, departments, or universities. UD presents other options and perspectives on access that will ultimately benefit all students, disabled and nondisabled.

Finally, being respectful of legitimate concerns around the time, effort, and energy required to alter course and instructional design in the context of UDL, it is perhaps important to consider UDL as an "upgrade" to teaching methods rather than a "rebuild." Faculty are encouraged to highlight application to the world beyond the classroom. Finally, UDL does not alter the fundamental concepts to be taught, but rather enables the continual evolution of teaching methods to respond to the needs and learning styles of the students. UDL, in that way, is true student-centered learning.

Inclusive teaching practices

Chapter Outline

Introduction 140
Principles of inclusion 140
Identifying implicit expectations and making them implicit 142
Approaches to developing inclusive teaching practices 144
Principles of effective accommodation 145
    Accommodations are individualized 145
    Individualized accommodations are flexible 145
    Accommodations must be developed and implemented in a timely manner 146
Conclusion 148

My science teachers were great, particularly during my primary and secondary education. Even without many resources in the Caribbean, they recognized someone who was hungry to learn (never mind the sight impairment) and did their best to nurture that. The best educators are ones who find a way to nurture that spark of interest within their students, and I admit – I've had some great educators over the years.

My teaching style today owes a lot to those educators. To be totally fair, it also owes a lot to the educators that I chose to not emulate. I learned from them what NOT to do…

From my best teachers I learned flexibility, creativity and innovation in teaching methods. I also observed how they did their best to be inclusive of me – both my accommodation needs and my learning styles – in their teaching, and the effort that they willingly expended on their students.

From my best teachers I learned the value of truly understanding the multiple purposes of an assessment, about the value of an assessment as a means to an end, as opposed to the end itself. I learned how to be flexible in designing my content assessments, fitting them to the needs and learning styles of my students.

I learned the value of authentic participation in learning the sciences. My best teachers never short-cut my involvement or engagement in the class, and worked with me as best they could to identify ways that I could learn in the context of my visual disability.

My best educators – TAs and faculty – in postsecondary, even at the graduate level – embodied the same traits, suggesting to me that there was nothing unique about primary and secondary education, or about teacher training. It was a set of characteristics that my best educators had that drew me to them. Fundamentally, it was who they were, and not how they were taught to teach (or not).

Creating a Culture of Accessibility in the Sciences. DOI: http://dx.doi.org/10.1016/B978-0-12-804037-9.00012-7

This was part of my "hidden curriculum" – knowing that I myself wanted to be an educator, that my aspiration was to be like my teachers, I watched, listened, learned. From the very best of my education in the sciences, I learned how to be a dynamic and inspirational educator myself.

Introduction

Enrollment in post-secondary STEM programs is becoming increasingly diverse. Administrators, faculty and staff in educational settings across many jurisdictions are legally required to provide an accessible environment for students with disabilities (see the discussion on the UN Convention on the Rights of Persons with Disabilities and other relevant accessibility legislation in chapter 2). To date, most of the literature addressing disability in post-secondary education in the sciences has focused on the creation of physically accessible science laboratories (for a review see Sukhai et al., 2014). While physical accessibility is an aspect of accessibility, it is important to realize that accessibility entails additional considerations.

An accessible learning environment requires educators to think critically about their course content and select teaching practices that foster learning for all students (Logan, 2009), as these items significantly affect a student's ability to participate fully in science, technology, engineering, and mathematics (STEM) programming. Inclusive teaching practices are selected or developed using a continuum of frameworks that include universal instructional design, differentiated instruction, and individualized accommodation. These frameworks can be applied to the teaching and learning process from the initial stages of course design to final course evaluations to improve accessibility for students with disabilities. This chapter outlines the principles of inclusion and effective accommodations, and illustrates how universal design, differentiated instruction, and individualized accommodations can contribute to the identification and removal of barriers for students with disabilities. We also suggest that educators need to be mindful of implicit curriculum expectations and informal learning processes that often occur outside structured labs and classes. Though these items are not often considered by educators when thinking about inclusive teaching practices, students with disabilities often encounter barriers to learning inherent in the "hidden curriculum" and interaction with peers required to achieve the essential expectations of a course.

Principles of inclusion

In a postsecondary educational context, inclusion can be understood as an *ongoing process*, where educators provide equal opportunities for students with diverse learning needs to be successful, through the identification and removal of barriers (UNESCO, 2005). When barriers are identified and removed, all students can *participate* (where participation refers to the quality of the student's experience in the course) and achieve the expectations of the curriculum (UNESCO, 2005). For science and technology disciplines, the student needs to be able to participate fully and authentically in different aspects of STEM education including labs, field work, tours, co-op or clinical placements, conferences (graduate students), and student groups or other extracurricular activities within the institution.

For graduate or professional program students, additional aspects of participation in STEM education also include clinical or research conferences and lab meetings, as well as seminars by visiting scholars. Full inclusion does not mean the student must be able to physically complete experiments in the lab in the same manner as his or her peers for whom a diagnosed disability does not create a functional limitation, but it does mean that learning activities are constructed respectfully and in light of the essential requirement(s) of the course, program, or discipline the activity was designed to address (see chapter 10: Essential requirements and academic accommodations in the sciences). Conceptualizing inclusion as a process where barriers are identified and then removed (UNESCO, 2005) means educators must *engage in ongoing data collection* to identify barriers to student achievement.

Assessment is the process of gathering information that accurately reflects how well a student has achieved the expectations of a course or program. The three types of assessment are assessment for learning, assessment as learning, and assessment of learning. Assessment *for* learning is ongoing assessment designed to give educators information to modify and differentiate teaching practices based on what the students need to be successful (OME, 2013). Assessment *as* learning is used by students to monitor their own learning and to make changes to the way they understand concepts or perform a skill (OME, 2013). This type of assessment can also be used by educators as assessment for learning. Assessment that is not part of a student's grade is referred to as formative assessment.

Finally, assessment *of* learning (summative assessment) is used to make judgments(s) about how well a student has achieved the expectations of a course, and often has little effect on learning (OME, 2013). Students should be provided with an opportunity to receive formative feedback addressing their achievement of the learning expectations prior to a summative assessment. Instructors should be mindful not to design assessments based on what is "easy" for the instructor. Instead assessments should be designed to allow the best opportunity to demonstrate their achievement of the essential requirements of the course. For example, multiple-choice assessment methodologies are easy for instructors to grade (e.g., using bubble cards and scanning equipment), but this method may not be the most accessible for all students or the most appropriate way to test a required competency.

A second principle of inclusion is for both students and educators to understand and for educators to develop (if they do not already exist) appropriate *essential requirements* for courses, as it is impossible to evaluate student achievement and plan for the removal of barriers if expectations are unclear (see chapter 10: Essential requirements and academic accommodations in the sciences). The essential requirements of the course, program, and discipline must be communicated explicitly and upfront to students and should be used in the construction of assessments. Furthermore, as discussed in chapter 10, Essential requirements and academic accommodations in the sciences, care should be taken to not conflate the *method* of assessment of the essential requirement(s) with the *requirement(s)* themselves. Thus, when planning assessments against the essential requirements of the course, program, or discipline, it becomes easier to design them from a universal access perspective and to develop and implement appropriate assessment accommodations for students with disabilities. This concept is particularly significant in the context of laboratory courses and lab-based examinations where a student may be graded on his or her ability to execute a timed lab activity in the absence of a written protocol. In such situations, if marks are assigned to the timing of the student's completion and his or her remembrance of the experimental protocol, students with disabilities may be penalized in both of those areas. If timing is critical to the experiment's success (and in some fields and instances, it is), then its assessment should be built into the measures of the experiment's outcomes, and the given timeline for performing the experiment and/or the expectation to memorize the procedure documented in the essential requirements for the course.

The speed and manner in which students construct meaning or develop technical proficiency are defined by their prior knowledge, experiences, physical, sensory and cognitive abilities, language skills, and their knowledge of and appreciation for the ideas of others. Inclusive education responds to this diversity by providing *flexibility* to the learners, so that all students have equal opportunity to make use of instruction. Examples of flexibility in an educational context may involve alternative methods to demonstrate their learning, multiple representations of a concept by an instructor, multiple ways to acquire knowledge, variation in the tolerance for error (variation in the amount of practice provided to students before final evaluation), and a variety of ways to engage with requisite material and skills.

Another important aspect of inclusive education in the postsecondary context is that *all students are expected to achieve the same essential requirements*. For students with disabilities, the role of specific, ongoing, and timely feedback around students' achievement of essential requirements is critical in order to continually refocus learning (e.g. the student could change the amount of time spent practicing a specific task or how the task is practiced), and identify barriers to achievement (Needs, 2002). Many students may not disclose a disability and/or a request for accommodation until they receive feedback that highlights unsatisfactory achievement of the expectations of a course (see chapter 6: Disclosure in the sciences). In many undergraduate STEM courses, students may be halfway or three-quarters through a course when they fail a midterm. Armed with feedback from the instructor about his or her unsuccessful performance academically, a student may request assistance. For some students, this may represent a crisis situation. In these situations, timely and ongoing assessment from instructors from the early stages of a course can help prevent crisis situations from developing in the first place.

Identifying implicit expectations and making them explicit

Students may still have difficulty achieving the essential requirements of a course, despite reasonable accommodations and the instructor's efforts to provide early and ongoing feedback to students, because the essential skills embedded in larger learning expectations are not explicit. The knowledge and skills educators intentionally teach to students that are identified on a course syllabus and communicated in lectures, field trips and labs are part of the explicit curriculum. However, the ability of a student to meet those overt expectations is often influenced by the student's access and knowledge of the "hidden curriculum."

This covert set of learning criteria consists of the unspoken, unwritten, or otherwise implicit academic, social, and cultural expectations that are communicated to students while they are in school. This information is often unacknowledged and/or unexamined by both students and educators, and yet, may be fundamental to achievement in the discipline. Science students, for example, are often expected to understand how to formulate testable hypotheses without direct instruction. Another example concerns the specific knowledge required to achieve writing proficiency in a specific discipline. Proficient academic writers understand *implicit, discipline specific* guidelines that are not taught directly, such as what counts as evidence in a discipline, how much evidence is required (Lakoff, 1990); how authority is constructed (Badenhorst et al., 2015); and how to present an argument in a genre (Elander, Harrington, Horton, Robinson, & Reddy, 2006). The "hidden" curriculum of academic writing may be of particular concern for upper-year undergraduate or graduate students who may have large amounts of their evaluation generated from written tasks. Educators need to identify and examine inherent expectations about how science and research writing is taught in order to give all students access to the curriculum.

In the health sciences (i.e. professional programs such as nursing), the hidden curriculum contains behavioural expectations around the development of a professional identity, leadership, ethics, and the ways of thinking and behaving professionally (Tsang, 2011). Students with disabilities that impact their ability to acquire, interpret and/or exchange information in social or other non-formal contexts may have particular difficulty with unpacking the expectations of the hidden curriculum relative to their peers for whom a diagnosed disability does not create a functional limitation. Examples of diagnosed disabilities which may impact social skills include learning disabilities, vision loss, autism spectrum disorders and hearing loss. As these students may break "social rules" without intent or even knowledge that they are doing so, they may require *explicit* instruction around professional ways of interacting that are not implicitly identified and/or directly taught. In some situations, students may be unsuccessful in these programs if they fail to meet expectations around professional behavior, even if they are academically proficient and/or competent technically. Thus, exposing the "hidden curriculum" around pragmatics and social skills is a significant aspect of an accessible education. Social skills also help cultivate relationships with peers and faculty. Therefore, students with accessibility needs in non-professional STEM programs also may require more explicit knowledge and teaching of social skills.

The hidden curriculum also refers to the values, beliefs, and often unintended lessons that shape students' learning (Balmer et al., 2009). Within the higher education context, students with disabilities are often discouraged from pursuing laboratory-based science programs (Hilliard, Dunston, McGlothin, & Duerstock, 2011; Miner et al., 2001; Moon et al., 2012). Students with disabilities may be aware of educators' diminished expectations for them as compared to their peers for whom a diagnosed disability does not create a functional limitation regarding achievement of course expectations. This discouragement may lower student motivation for learning, and/ or manifest in the educators effort to teach these students effectively. These attitudes may further negatively affect the overall culture of the classroom if they are communicated to the student's peers. Educators need to possess the attitude that all students can learn, progress, and achieve such that they have consistent and high expectations for all students. Finally, an important principle of inclusion is that it fosters open, ongoing and effective collaboration between the student, the student's instructors/supervisors, accessibility services staff, and administrators around the student's academic progress and any accommodation(s) required (Muller, 2006). A student's peers may also be important collaborators as peers may be essential to the implementation of accommodations in the classroom and/or for group work. For a more detailed discussion, see chapter 15: The student in a leadership, mentorship, and supervision role.

Approaches to developing inclusive teaching practices

Inclusive teaching practices refer to the structures and practices used to design courses, deliver courses and assess student achievement. These techniques can be identified through the application of three approaches that outline principles to make learning accessible for diverse groups of students. Each of these approaches specifies different practices to make learning accessible to students with disabilities. A proactive approach is universal design for learning (UDL) as discussed in the previous chapter and elaborated on here in subsequent sections. Two reactive approaches to inclusionary teaching include differentiated instruction and individualized accommodation. Educators who proactively design their courses with UDL principles (chapter 11) in mind may be able to minimize but not fully eliminate the need for individualized accommodations for students with disabilities (Shaw, 2011). All three approaches can be used together to create access to STEM programs for all students.

While UDL identifies broad principles for planning instruction for diverse groups of students, differentiated instruction allows educators to address specific skills and difficulties *as they arise* over the course of instruction (Raynal & Rieunier, 1998). Hence, UDL is about planning ahead of time to minimize the need for individual accommodations, whereas differentiated instruction is a framework that guides educators to respond to learning needs as they arise. However, it is important to note that differentiated instruction does not mean doing something different for each student in the class (Tomlinson, 1999), and thus does not include individualized accommodations. Since differentiated instruction responds to learning gaps as they arise, it is informed by ongoing assessment of students. Ongoing assessment does not have to be tedious for the instructor. An "exit" card with a response to an assessment question at the end of a class is an example of assessment data that is easily collected. Differentiated instruction may occur at the beginning of, or during a "unit" of study. At the beginning of a "unit" of learning, instructors may change an aspect of the instruction for different groups of students based *on prior assessment* (e.g. pre-test, or survey) of the students' level of background knowledge or prior achievement of the learning requirement. Or, during the course of instruction on a particular topic, an instructor should also use *assessment for learning* to identify students who have not yet attained a learning skill while other students have, and differentiate element(s) of instruction for *that group* of students accordingly.

Within a differentiated learning model, there are four main elements of instruction that can be varied to respond to student needs. These include the *content* of the learning, the *process* of learning (including materials and flexible groupings), the *products* of learning (e.g., assessment of learning), and the *environment* of learning (Tomlinson, 1999). In a specific laboratory activity, student groupings would be informed by the instructor's assessment from a previous activity. One group might be ready to move ahead independently with little instruction, another group might need more structured guidelines (more detailed written instructions), and a third group might require additional demonstrations or explanations from the instructor. Educators may assign different problems, different materials, different starting points, or different ways of demonstrating learning to different groups based on ongoing assessment. Since differentiated instruction does not differentiate for individuals, it is less work for instructors than individualized accommodations.

Principles of effective accommodation

Accommodations are individualized

Even when a course is designed and implemented with the principles of universal instructional design and differentiated instruction, differential treatment may sometimes be required to provide individual students with an equal opportunity to demonstrate the essential requirements of a course.

Effective accommodations respond to the unique circumstances of each individual. Barriers encountered in postsecondary education by students with disabilities are shaped not only by the nature of their disability, but also by the interaction of their disability with their learning environment (e.g., physical environments, institutional procedures and policies, program-specific essential requirements, learning activities in a course of study, etc.). For this reason, accommodations need to be selected based on the unique needs of the individual, and not solely based on disability category or other generalizations, or they will not be effective (OHRC, 2004).

Individualized accommodations are flexible

In STEM programs, students may encounter a variety of learning environments and therefore will have a range of learning needs. These may include labs, fieldwork, lectures, group works, conferences, oral examinations (e.g., a thesis defense), and lab meetings. An individual student may have different needs in each of these environments due to the physical surroundings, the nature of his or her disability (e.g., mental health and/or episodic disabilities), the nature of the physical task and/or communication (including the modes used to present information), and the interaction of all of these factors. Flexible accommodations respond to the *evolution of barriers* that may be encountered by a student when he or she is in school. At the most fundamental level, flexibility with respect to accommodations means a student's accommodation plan may need to be continually examined and modified as he or she progresses in a program.

Accommodations must be developed and implemented in a timely manner

Though it may be challenging for faculty, staff and students to find time to develop, implement, review, revise and then implement revised accommodations, effective accommodations must be developed and implemented in a timely manner. When a need for *revision* to an accommodation plan arises, it is usually either due to an evolved barrier or an accommodation has been found to be ineffective at eliminating a previously identified barrier. Until the accommodation is put into place, the student will have at least more limited access to instruction, and may not be able to acquire the necessary information and/or skills needed in time for a summative assessment. Until a student's accommodation(s) are in place, consideration must be given to the inaccessible content of the course and the schedule of summative assessments so the students' grades are not penalized during the period in which they are waiting for accommodations to be implemented. In a situation where the schedule of summative assessments may need to be revised to accommodate for the time the student hasn't had an accommodation in place, the student is still responsible for demonstrating the requisite knowledge and/or skills in a course. That is, effective accommodations *do not compromise or modify the essential requirements* students are expected to meet by the end of a course.

One of the reasons accommodations may not be implemented in a timely manner is that sometimes *creative accommodation solutions* are needed for students with disabilities to meet the essential requirements of a course. Students with sensory disabilities in particular may require novel and creative accommodations to participate fully in STEM courses, since in these types of programs, traditional classroom activities are combined with experiential activities that frequently engage a student's senses (particularly vision and hearing) and require students to access information presented in graphs, charts, videos, or demonstrations.

Technological solutions represent an important aspect of accessibility for these students, and may only be realized through innovative ideas arrived at through collaborative efforts. In addition, as the number of students with disabilities in STEM (particularly sensory disabilities) is low, there are few precedents to provide direction to the institution on how to handle accommodation challenges in STEM programs. Accessibility advisors may not be aware of the exact nature of a particular experiment and/or the essential requirements of the course, faculty may be unfamiliar with specific challenges posed by the interaction of an environment with a disability, and everyone collectively may not be aware of potential technological solutions that may be part of an accommodation. Thus developing novel and creative solutions to accommodation challenges is best approached by a team comprised of the student, faculty, and disability services staff. Since accommodations do not excuse students from demonstrating the required essential requirements, evaluating the appropriateness of a novel accommodation dictates the need to identify what those essential requirements are. It is not recommended that judgment of this should not be left to individual faculty members. For this reason, administrators need to be involved in accommodation processes and decisions.

Effective accommodations allow students *to participate authentically in labs, group work, and all learning activities indicated in a course syllabus.* When a student has vision loss, and cannot participate in the dissection in a biology class, simply allowing the student to observe the dissection is not an appropriate accommodation, as the student cannot visually take in the intricate details of what is occurring around them. If the essential requirement of the activity is to learn the anatomical parts of the frog, then having the student with vision loss "observe" his or her peers changes the student's ability to achieve the essential requirement. Even if the student participates in a computer-simulated dissection, this is still not necessarily an appropriate accommodation if the software does not capture the 3D of the frog's anatomy, which would be inherent in the actual dissection of the frog by the student's nondisabled peers. An appropriate accommodation must expose students to the same type of information their peers (for whom a diagnosed disability does not create a functional constraint) are exposed to, to give the student with the accommodation equal opportunity to achieve the essential requirements the activity is designed to address.

In another example, if the essential requirement of a molecular biology lab is to understand how to troubleshoot results appropriately by making changes to the procedure, an appropriate accommodation for a student unable to physically perform the lab (i.e. perform gel electrophoresis) is a lab assistant assigned to that student. In the case of troubleshooting, the student asks the lab assistant a series of very specific questions based on a thorough knowledge of the technique in order to fully understand the results of the experiment and determine an appropriate modification to the procedure or explanation for those results.

Group work is a popular pedagogical strategy used by faculty in many post-secondary STEM programs. Successful communication with peers may even be an essential competency for some professional programs. While we know the benefits of peer-to-peer learning (Tierny, Grossman, & Resch, 1995), it does pose some unique challenges for students with disabilities. For graduate students, peer collaborations may significantly facilitate academic achievement, as students may form writing groups and publish together.

Consideration of the components of group work will facilitate understanding of the challenges that co-operative learning provides for students with disabilities. Group work requires participants to perform quick reading and mastery of essential facts when distributed or assigned in the classroom; remaining focused and continually investing energy until a group decision is reached; taking in, comprehending and retaining others' ideas; contributing appropriately to the discussion; the ability to articulate a position in relation to the course of action tentatively favored by the group; recording ideas and the developing decision of the group and presenting the final decision to the class. Students with disabilities that impact their communication (receptive or expressive, written or oral) or their ability to read social cues may have particular difficulty with group work. Sustaining focus and attention would be difficult for certain students with ADD/ADHD, depression, or a chronic health issue. Students with Asperger's or autism spectrum disorders may be overwhelmed by too much sensory information, and have difficulty processing verbal information and responding to others. Students with disabilities may also struggle with working in groups due to attitudinal barriers among peers, and/or a lack of awareness on the part of the student with the disability of their specific need(s) for accommodation in group situations. To facilitate the development of appropriate strategies for working in groups, students with disabilities and those that work with these students need to understand the specific skills required for successful group work, the specific course requirements around the group work, and how the student's disability will interact with these expectations. Once a student's accommodation(s) for working in groups are identified, the student's peers need to understand the specific disability related issue(s), and the specific aspects of the accommodation. This may require the student and faculty member to educate peers. One way this may be achieved is through a group contract. A contract might list group communication norms. An example of a communication behavior that might be supportive is that if the student needs extra time to process what is being said, the group might need to be aware of their discomfort with silence, in order to give the student with the disability extra time to respond. A contract might even list a specific amount of extra time the group would wait (silently) while they wait for a response before making a decision. Faculty need to clarify the expectations for the group work with all students, be approachable, and willing to assist with both the development and implementation of specific accommodations around peer collaboration.

Conclusion

Faculty and staff occupy important roles in facilitating the academic achievement of students with disabilities (Zhang, et al., 2010). By using inclusive teaching frameworks, educators will be positioned to develop pedagogical practices that respond to learner diversity. Institutions need to offer authentic opportunities for faculty to learn about and apply these approaches to all stages of their professional teaching practice; from the development of essential course requirements to summative evaluations. Administrators and faculty also need to uncover the expectations inherent in the "hidden curriculum," and explicitly teach these expectations to students. Furthermore, faculty need to recognize the types of "hidden" barriers inherent in group work and think more broadly when they consider the types of learning activities students may participate in outside the classroom but within the institution that need to be inclusive.

Part V

Students as Mentees, Trainees, and Leaders

Faculty mentorship of students with disabilities in the sciences

Chapter Outline

Introduction   153
The importance of mentorship in STEM   154
What is mentorship?   155
Mentors can be people with or without disabilities   156
Forms of mentorship   156
   In-person (one-on-one) mentorship   157
   Online mentorship   157
   Senior/faculty mentorship   157
   Peer mentorship   158
Qualities of a good mentor   158
   A good mentor is proactive   158
   A good mentor is responsive   159
   A good mentor is open-minded   159
   A good mentor is creative   159
Benefits of becoming a mentor   160
Selecting a good mentor   160
Conclusion   161

It was overwhelming during my undergraduate degree, and then when I began my Master's, to know that there was no one who had been through my program before me who had a disability. No one that I knew who was a scientist with a disability. There was no one I could turn to who could help me through the program – no one who could offer advice, who could tell me of their own experiences, who had any tips and tricks to get through the hard days and long nights. I felt I had to do it all myself.

I didn't feel like I could turn to the faculty either, as an undergraduate student; I just didn't feel confident enough to approach any of them and ask for their advice. I also felt – for a variety of reasons – that I was not welcomed as a student in my program. Whether that was a real thing, or just my perception was not relevant, as, regardless, it had a real impact on how I approached my undergraduate education – and to a lesser extent my Master's as well.

When I started my doctorate, I took a more intentional approach to finding mentors – it became easier to do so in a larger department

Creating a Culture of Accessibility in the Sciences. DOI: http://dx.doi.org/10.1016/B978-0-12-804037-9.00013-9

with more collaborators, and as I was able to attend national and international conferences, I was able to identify people across the continent I could approach to be my mentors. During this time, one of my faculty mentors said something wonderfully affirming to me – he said, "You don't need your eyes to do science." Finally, I had confirmation that my childhood perceptions of what science was, were accurate! I return to those words any time I feel like I'm struggling as a scientist, and when I doubt myself and my abilities.

Even as a doctoral stream student, though, I could find no person with a disability who'd attempted what I had done. In many ways, I was a trailblazer – and indeed, am now acknowledged as such by students with disabilities who have come after me. I did not do this for any other reason than I had to – and even so, I fell into it without conscious forethought.

During my doctoral studies, I began to make a conscious effort to find young professionals with disabilities – outside the sciences, because there were none that I could find at that time in the sciences – whom I could talk to. I was interested in learning about technology and the use of technology in accommodation; I was interested in learning about others' journeys through their professions, and in drawing out the lessons I could learn from these different experiences.

These peer mentorship experiences complemented nicely the mentorship I was able to get from collaborators and faculty within my department, as well as from colleagues at both a national and international level, and once I saw the benefits of the time invested in these relationships, I took the effort to maintain and expand my networks as I entered my postdoctoral training and the roles which I currently have. I believe that I can never have too many mentors, and that there is always something new to learn.

Not everyone is suitable to being a good mentor, though – I have met any number of people I have discounted from being potential mentors – they were not understanding of my needs and interests; they were dismissive of my disability or my career goals; their perspectives were too far removed from my philosophy to learn much; or, they were genuinely not interested or too committed to spend the time to engage with me. Whatever the reason, I grew to recognize that mentorship is not something that everyone will actively be interested in, and that it is worth it to be selective in order to identify the best and most appropriate mentors.

Now, with my training behind me, and having settled into my own career, I find myself in demand as a mentor. I have done something others have not yet succeeded at – recognizing that, I feel the obligation to give something back to my alma mater, and the trainee community at large – particularly for students with disabilities. Indeed, this whole book is, to some degree, an expression of that obligation, on a much bigger canvas. In mentoring others – be they students and trainees (with or without disabilities) in my discipline, or peers with disabilities outside my field of study – I try to emulate the very best characteristics of the mentors I have been privileged to have, and in so doing pay their service forward to the next generation.

Introduction

Faculty have an important mentorship role to play with students with disabilities taking lab-based science courses and programs. This is particularly true, and much more self-evident, for graduate programs of study, where there is a direct one-on-one relationship between the student and the supervisor, and where the mentorship relationship is formalized to a significant degree (Herrera, Grossman, Kauh, Feldman, McMaken, and Jucovy, 2007). However, even at the undergraduate level, anecdotal evidence suggests that students with disabilities will succeed in the sciences generally, and in lab courses specifically, with faculty who are more willing to be engaged and take an interest in the student's success. Conversely, faculty who present as indifferent, unsure, discriminatory, or outright hostile will negatively affect a student's success to the point that he or she may simply drop the course or leave the program completely (Pence, Workman, & Riecke, 2003). Some students have noted anecdotally that their best instructors were often young and willing to engage directly and in a "hands-on" manner with accommodations and adaptive technology to accomplish something new. The best instructors were also honest about not having a solution on hand, while being willing to work with students to figure something out.

Students with disabilities often don't explicitly seek formal mentors at the undergraduate level. Given the relative dearth of role models for these students in the sciences, mentorship is usually not a priority. If a faculty member demonstrates a willingness to engage with the student and shows an interest in his or her professional development, an informal mentorship relationship may take root. In these relationships, students are interested in:

- Identifying a faculty member who is open-minded about the inclusion of disability in the sciences, and who demonstrates this open-mindedness in the course of his or her interaction with the student;
- A champion or advocate who is able to help them navigate the discipline and laboratory setting, as well as the interface with the academic environment, in a way that the disability services staff may not be able to;
- A faculty member who demonstrates creativity and a willingness to critically think about the interface between disability and the lab-based environment; and
- A faculty member who is relatable, approachable, and responsive to student interaction (Sukhai et al., 2014).

It is worth noting that a student who has identified a faculty member as a potential mentor is likely to allow the relationship to develop naturally, without actively seeking its formalization. In many instances, it may be up to the faculty member to recognize that a mentorship relationship is evolving, and to act accordingly. It is also possible that a student may gravitate toward a faculty member with whom he or she has had some prior interaction, but who is not currently involved in his or her courses.

At the graduate level, because of the importance of the graduate thesis supervisor and the relationships between the student, supervisor, and thesis advisory committee members, mentorship relationships are likely to evolve naturally. In this context, students with disabilities may not be fully aware of the differences between undergraduate and graduate training (e.g., course-based versus research and one-on-one/small-group interaction), and may approach these relationships as they would have in undergraduate settings. It is critical for students, both with and without disabilities, to understand the central importance of mentorship in science education (Metros & Yang, 2006). Regardless, graduate students with disabilities need to look for the same characteristics in a good mentor as they did at the undergraduate level.

Finally, a good faculty mentor has several qualities, particularly with respect to mentorship of students with disabilities, which include being:

- *Proactive*: Faculty members should be willing to reach out to and engage the student on his or her own terms, as opposed to waiting for the student to come to them with a crisis.
- *Responsive*: Mentors must respond to student engagement in a timely manner.
- *Open-minded*: Faculty members should demonstrate an inclusive mindset with respect to the involvement of students with disabilities in the sciences and science labs.
- *Creative*: Mentors who demonstrate creativity in thinking about issues faced by their mentees in the context of their disability, and are more willing to think critically about adapting the essential requirements of the program to the student, are more likely to be successful (Sukhai et al., 2014).
- *Calm and approachable*: Mentors who think they need to have "all of the answers" or "fix everything" are setting themselves and their mentees up for additional challenges. A mentor who can be easily approached and who can have calm open discussions will allow the student to have a voice in the discussion. Mentors are not expected to "fix" challenges, but rather to support the student as they navigate either traditional or nontraditional obstacles in nonconventional ways.

Subsequent sections of this chapter will expand on these issues and present best practices and solutions for mentorship in science, technology, engineering, and mathematics (STEM) for both students and their mentors.

The importance of mentorship in STEM

Mentorship is a key part of success in many STEM fields. This is true for a number of structural reasons including the fact that STEM disciplines are complex and multifaceted, STEM careers are diverse and can be difficult to navigate without "inside knowledge," and mentorship helps trainees navigate this complexity in STEM. Today, mentorship in STEM can start as early as the undergraduate program, and even, with the right cultivation on the part of the student and the science educator, in high school. Indeed, training in STEM allows for "facilitated" mentorship—in other words, opportunities to engage with mentors are often included for students as part of their program.

These mentors may be faculty (formal mentorship through supervision or membership on advisory committees) or peers (informal mentorship in the research environment) (National Science Foundation, 2013). Finally, it is worth noting that mentorship and supervision are two very different and distinct things—in other words, your supervisor can be your mentor, but your mentor does not have to necessarily be your supervisor. This is an important distinction, and one we will return to in a subsequent chapter.

For a variety of reasons (e.g., disclosure, lack of student engagement, lack of appropriate information or education, etc.), students with disabilities often do not have access to mentors. Since mentors are often role models who embody characteristics a mentee can learn from and emulate, the increased difficulty in identifying appropriate mentors can be a significant barrier to students with disabilities and their success in STEM. The remainder of this chapter will address issues of mentorship in STEM from both the mentor and mentee perspective and identify leading practices that may foster increased involvement in mentorship relationships for students with disabilities.

What is mentorship?

Before introducing best practices, it is important to introduce some of the key concepts that inform how we think about mentorship in STEM, and why it is so important to the success of students with disabilities.

Mentorship is a relationship formed between two people with the goal of sharing knowledge and expertise. Mentorship is often thought of as guidance of a trainee by a more experienced individual, but it does not have to be limited to this type of relationship. Mentorship is considered to be an opportunity to be supported by someone with a defined scope of expertise that would be of benefit to the student.

Mentorship can be a formal relationship with written goals and scheduled meeting times, or it can be as informal as an occasional chat or email exchange. At one extreme, mentorship can be structured as part of an independent program or incorporated into a course or discipline program of study (Cavell, DuBois, Karcher, Keller, & Rhodes, 2009). At the other extreme, mentorship can also be very informal and unstructured, and be at the discretion of the two individuals (the mentee and the mentor). This type of mentorship is not associated with any formal requirement or programming, nor with a specific organizing body, program of study, discipline, or career choice (Metros & Yang, 2006).

Mentorship is a relationship between at least two individuals. The mentor is a coach, guide, tutor, facilitator, counselor, and trusted advisor. A mentor is someone willing to spend his or her time and expertise to guide the development of another person (MENTOR, 2009). Mentorship can be "multilevel"—i.e., mentors can be higher-level students or trainees, faculty, supervisors, advisors, collaborators, etc. Furthermore, mentors do not have to be from the same field as the mentee. The mentee is the individual who is receiving support and guidance from the mentor (e.g., student, protégé, apprentice, eager learner) (Rhodes & DuBois, 2006). A mentee is someone who wants to learn from someone who knows and seeks his or her valuable advice in order to grow personally and/or professionally.

Mentors can be people with or without disabilities

There are myriad benefits to having multiple mentors, both with and without disabilities. Students should strive, where possible, to engage with both types of mentors. Students with disabilities may be more inclined to identify—and identify with—mentors with disabilities themselves, but given the low participation rates for persons with disabilities in STEM, these mentors can be difficult to find. However, mentors with disabilities are important and offer a unique perspective, as they can act as role models and can answer questions around disability, accommodation, and participation in STEM. Mentors with disabilities may also be able to offer their mentees the benefit of their networks, but may not be in the same field as their mentees.

Mentors without disabilities, on the other hand, can help their mentees break down barriers to their participation in the field by helping dispel stereotypes about persons with disabilities in STEM. Furthermore, mentors without disabilities may be more directly connected to networks in the same field of study as the student, which may be of direct benefit to the student in his or her career and professional development. Mentors without disabilities may also encourage their mentees to stretch their thinking and redefine their perception of their limitations or boundaries in ways that a mentor with a disability may not be able to do, because of their own experiences.

Mentors without disabilities, particularly those in the same field as students, may also help counter the "gatekeeper function" previously discussed (see chapter 3: Barriers faced by students with disabilities in science laboratory and practical space settings and chapter 11: Universal design for learning) by serving as advocate and guide in situations where other faculty and professional accreditation bodies may raise objections to students' participation in STEM. This advocacy role could be taken on independently of the mentee's involvement, although it should be done with his or her knowledge. A mentor without a disability can also role model the benefits of working with students with disabilities in their field to their colleagues, department chairs, and senior administrators, and in so doing can begin to counter the systemic issues associated with the "gatekeeper function" in higher education.

Mentors with different backgrounds and abilities can bring multiple perspectives to the mentee, which can be beneficial to the student's growth and development in STEM, as well as later in his or her career. Students should therefore be encouraged to seek out mentors from diverse backgrounds, both with and without disabilities, in order to broaden their perspectives about STEM and their ability to succeed in their chosen field.

Forms of mentorship

Mentorship comes in a variety of forms, depending on availability of the student and mentor, resources, location, and the specific needs of both the mentee and mentor. Some forms of mentorship include in-person (one-on-one) mentoring; online or e-mentoring; mentorship between two peers; and mentorship between a student and their faculty.

In-person (one-on-one) mentorship

In this form of mentorship, students can get to know the mentor personally, and build a stronger connection with him or her, because they are able to meet in person for longer periods and/or more frequently. The mentor–mentee relationship is qualitatively different in person, as the depth of conversations is often greater in this form of mentorship, and the in-person connection can lead to additional practical components (e.g., demonstration of lab-based activities or technology solutions) that other forms of mentorship may lack. However, there are also some disadvantages to in-person mentorship. For example, the student may be limited by location, transportation, or mobility in terms of the available pool of mentors. In smaller cities or rural areas, there may not be any mentors available. Additionally, it may be harder to find the time and flexibility to meet in-person regularly, particularly for long conversations. In this case, students may need to connect with mentors through phone or online, and reserve longer in-person meetings for when both parties are available.

Online mentorship

Although online mentorship presents challenges in building strong and deep mentor–mentee relationships, it is the most effective method of finding mentors. Online mentorship is not limited by location, transportation, or ability, and it is not limited by national or international boundaries; it is possible to recruit mentors from around the world. Meeting online opens doors to a broader pool of networking opportunities, both nationally and internationally. It may also be easier to identify a mentor from a different geographic region who more closely aligns with the student's own field and interests through this approach, as opposed to locally and in-person. However, online mentorship or e-mentorship also has some disadvantages. For example, there is less opportunity to get to know the mentor, or to have more "hand-over-hand" assistance and practical demonstrations of tools and resources, and there is the potential for more difficulty navigating the technologies required for online connections.

Senior/faculty mentorship

Establishing a mentorship relationship with faculty and senior members of the academy or in the employment sector is important for students with disabilities. These mentors provide the student access to a wealth of lived and mentorship experience, to networks and connections in STEM, and can open doors within the field. Senior mentors can also provide more academic and career advice. However, it may be harder for students to relate to these mentors, because of age differences, and these mentors may not have disabilities, making it more difficult for mentees to disclose their needs and concerns around disability and integration into STEM. Furthermore, these mentors may be busier, and have less flexibility and time to work with mentees. The student's goals with respect to the mentorship relationship (i.e., what they hope to achieve with the mentor), as well as how well the student has worked to ensure his or her goals are met, will determine how successful the relationship with the mentor will be.

Peer mentorship

In the case of peer mentorship, the peer mentor may be in the same age range as the mentee but is usually more experienced in one or more areas. Thus, for example, a peer mentor could be a graduate student if the mentee were an undergraduate student. The mentee may be more comfortable sharing his or her story and concerns with a peer, and may find it easier to ask the peer questions he or she would be reluctant to ask a mentor who may be in a potential supervisory or academic role. A peer mentor may be easier to relate to for the mentee, because he or she is in the same stage of life and has gone through similar experiences. Finally, a peer mentor may be easier to access and have more time or flexibility for mentorship than a more established mentor. However, peer mentors may not have a lot of expertise, life experience, mentorship experience, or connections within the field—thus to take advantage of a peer mentorship situation, the mentee needs to have a specific understanding of what he or she is hoping to accomplish with this relationship. From a peer mentor's perspective, there are benefits to serving in this role as well, as peer mentorship can be a personal growth and development experience for the mentor, and may be a valuable and needed component for the mentor's career advancement. Additionally, acting as a mentor can allow peers to develop needed leadership skills that will serve her or him well in future career and educational pursuits.

Qualities of a good mentor

There are several key considerations in becoming a good mentor, and several qualities that characterize good mentors in STEM. These include being proactive, responsive, open-minded, and creative. These four characteristics are necessary to ensure the student has a positive and productive mentorship relationship. As the mentor plays a crucial role in the success of the mentorship relationship, it is incumbent upon the mentor to nurture these characteristics within themselves.

A good mentor is proactive

Good mentors take steps to understand and engage with the their mentees; in particular, good mentors get to know their mentees' concerns, needs, and goals. This depth of knowledge will enable the mentor to recognize if and when the mentee may be in a crisis situation, and to offer sensitive, timely, and appropriate interventions. A good mentor demonstrates his or her proactiveness by looking for opportunities for mentee growth and engagement in the field, and by presenting these to the mentee for consideration. Mentors may also need to advocate for mentees with disabilities, especially in situations where the mentee may be working in a context (laboratory, discipline, etc.) where there haven't been prior students with disabilities. Ideally, such advocacy is undertaken with the mentee's knowledge and consent, particularly in situations where the mentor cannot engage the mentee as part of the advocacy effort. Finally, a good mentor foresees potential barriers the mentee may encounter in his or her studies and field, helps the mentee identify these, and offers advice in planning how to overcome these challenges.

A good mentor is responsive

At its core, effective mentorship is a volunteer commitment, distinguishing it from the conflation of mentorship and supervision that happens within academia (more on this later). Since this is the case, it is important that a mentor recognize the need to respond to his or her mentees in a timely manner, particularly given the nature of multiple commitments and the potential for multiple mentees who may need his or her advice and mentorship. A responsive mentor is one who engages with the mentees by reaching out regularly, offering advice and help, as well as opportunities to meet to discuss issues the mentee may be facing. It is ineffective for a mentor to respond to a mentee's request after the issue has been resolved.

A good mentor is open-minded

A good mentor is open to exploring new and creative ways to performing tasks and achieving goals with the mentee. A good mentor is also open to the participation of the mentee in his or her chosen field, and is willing to work with the mentee and support him or her in the accomplishment of that goal. Good mentors recognize that it is not their role to discourage or play gatekeeper to the mentee, but that it is their role to model, by their behavior, a positive and inclusive attitude with respect to his or her mentee's participation in STEM to colleagues and supervisors.

A good mentor is creative

As we previously noted, mentors are more successful when they demonstrate creativity in addressing issues faced by mentees with disabilities, and when they are willing to think critically about adapting the essential requirements of the program to the student. This is an important point and bears repetition. Creativity can be demonstrated by the mentor in a number of distinct ways. For example, the mentor can be of assistance in identifying unconventional accommodation solutions in the research or laboratory environment. The mentee, if he or she is new to the field, may look to the mentor for creative solutions and application of ideas to foster his or her participation in the discipline. The mentor may also be able to exercise creativity in helping the student to think through integration challenges and problem solve leading practices to overcome them. More senior mentors who know the department's or institution's policies and practices may be able to brainstorm creative solutions with the mentee in the funding or provision of accommodations in the STEM environment. Finally, mentors can also think creatively about the mentorship process itself, in order to identify solutions that would best fit the mentee's needs. This could include novel or alternative forms of mentorship (e.g., volunteer placements or job shadowing), group mentorship, or other ways to foster the mentee's interests and develop their experiences.

Benefits of becoming a mentor

There are several benefits, both internal and external, to becoming a good mentor. First, mentors find themselves with opportunities to build new and interesting relationships, both with prospective mentees as well as others involved in the mentorship process. Mentors get to expand and develop their mentorship skills, which is particularly advantageous to mentors with disabilities and younger mentors who may not have had many prior opportunities. Additionally, mentorship provides the opportunity to educate not only your mentee in the STEM field but also to educate other faculty, administrators, and staff within the department or institution. Mentors may also serve as leaders, or pioneers, as others within the institution become aware of their efforts and strive to emulate them.

Mentors can have increased opportunities to connect with others with similar experiences in the disability community, and can also gain access to valuable resources. Faculty may have the opportunity to gain knowledge of and experience with disability-specific accommodations in STEM, which may aid them when they are instructing and mentoring future students with disabilities. Finally, mentorship provides an opportunity to share knowledge and skills with a student.

Selecting a good mentor

It is critical for prospective mentees to think carefully about what they are seeking in a mentorship relationship, what they require in turn from a mentor, as well as how they should go about identifying prospective mentors (National Educational Association of Disabled Students, 2010). Creating and understanding one's goals for the mentorship relationship and linking this to the role the mentor will fulfill is a fundamental step in this process. Defining this carefully enables the mentee to determine what type of relationship is likely to work best, as well as whether the mentor ought to be a peer, educator, or more senior faculty member. It is also important for a prospective mentee to identify a mentor who is open-minded about the inclusion of disability in the sciences, and who demonstrates this open-mindedness in his or her daily interactions. Since different mentors may fulfill different needs for the mentee, it is important to also consider what specific needs should be addressed in each mentorship relationship. It is important for the prospective mentee to "screen" potential mentors in order to find the best fit, which can be done through informational interviews and short conversations. It is also important to be realistic about one's mentorship needs, particularly in terms of how the mentorship relationship should be implemented (e.g., in person vs online) and how it should evolve. In evaluating potential mentorship candidates, it is important for the prospective mentee to consider for themselves whether they wish mentors who are:

- A champion or advocate who is able to help students navigate the discipline and laboratory setting, as well as interact with the academic environment, in ways that staff from the disability services office may not be able to. Staff from disability services offices frequently do not have experience working in the sciences and therefore do not have the tools and experience to effectively guide students.

- Someone who demonstrates creativity and willingness to critically think about the interface between disability and the lab-based environment.
- Someone who is willing to have open and frank conversations with students with disabilities. A mentor who is comfortable working with students with disabilities is more capable of having constructive and meaningful conversations.
- Someone who is relatable, approachable, and responsive to student interaction.

There are many places that one can go to in order to identify and begin to research prospective mentors. These include departmental and institutional websites for your school, as well as several online STEM resources, such as sciencecareers.org, the DO-IT project at the University of Washington, and disability-specific agencies such as the National Educational Association of Disabled Students (Canada) and the American Foundation for the Blind's CareerConnect resource (United States). Prospective mentors can also be found in person, through approaching faculty members, peers, or service providers locally. They can also be found through attendance at regional, national, and international conferences.

Conclusion

Mentorship is the guidance of a trainee (mentee) by a more experienced individual (mentor) and provides opportunities to be supported by someone with a defined scope of expertise. As noted mentorship and supervision are two very different and distinct things—i.e., your supervisor can be your mentor, but your mentor does not have to necessarily be your supervisor (this will be discussed in the following chapter). In this chapter, we discussed several forms of mentorship, their characteristics, limitations, and benefits, and how each may benefit the student.

Mentorship can occur either in-person or online, and this is dependent on geographical location, availability, and mobility constraints by either the mentor or mentee. A mentee may receive mentorship from peers, faculty, or from an individual already in the field. Students with disabilities often do not have access to the mentors they need, which makes it harder for them to make informed career choices, thus it is important for students to think carefully about what qualities they are seeking in a potential mentor to make the process of finding one easier. Students should strive, where possible, to engage with both types of mentors, those with a disability and those without. Students with disabilities may be more inclined to seek out mentors with disabilities. However, as there are not many people with disabilities in STEM fields, these mentors can be difficult to find. Mentors with disabilities are important and offer a unique perspective, as they can act as role models and answer questions around disability-related subjects, accommodation, and participation in STEM. Conversely, mentors without disabilities can help their mentees break down barriers to their participation in the field by helping to break down barriers to accessing careers in the STEM field.

Choosing a mentor is not an exact science. It is a process that requires people skills, experience, and good interviewing skills. These are qualities undergraduate (and all too often graduate) students lack by virtue of inexperience. Sometimes factors such as compatibility and differences in long-term goals make for a poor match between mentor and mentee. A successful relationship is based on common goals and mutual respect. If a mentor is more interested in being involved in the field of disability advocacy rather than in his or her mentee's specific educational or career goals, the mentee is at an increased chance for failure. For instance, if a student is interested in participating in "bench work" but the mentor is more interested in the mentee contributing to educational research, then the lack of commonality of goals may become a problem and a source of friction. Students must also learn to discern good relationships from less than compatible ones and to expect to have bumps along the road. Occasional failure serves as a valuable lesson, and it is important to develop the skills necessary to cope with situations that do not develop as planned.

Faculty supervision of students with disabilities in the sciences

Chapter Outline

Introduction   165
The foundation of the student–supervisor relationship   165
Disability, disclosure, and the student–supervisor relationship   166
Quality of the student–supervisor relationship   167
Deterioration in the student–supervisor relationship   168
Clarifying expectations in the student–supervisor relationship   168
Students in crisis   169
Supervisor's knowledge of and/or willingness to participate in disability-related processes   170
The role of the supervisor's knowledge of the interface between essential requirements and academic accommodations   171
At the interface of research integrity and accommodations: authorship issues   171
Supervisors may assist students with academic and social integration   172
Boundary issues   173
Funding issues   173
Delegated supervision   174
Conclusion   174

In every interview for graduate school admission, there would come a point where I would flummox my prospective supervisor. It was when I chose to disclose my visual disability. I never chose to hide this – I work in biomedical research, and for the last fifteen years, I have done so in a hospital setting, surrounded by trained medical professionals. I don't think I could have hidden my sight impairment if I'd tried. For those of us who have some form of a "visible" disability, not having to worry about disclosure is both a freedom and a curse – but it's still pretty stressful to have to bring it up in an interview setting.

I did it routinely, however, because the reaction from across the interview table usually told me something about the quality of the person I was interviewing, and whether or not I wanted to really work for them as an individual (to be fair, I only did it during interviews for graduate and postdoctoral positions with people I didn't know – there was no need to disclose with people I did know previously, particularly if they were trying to recruit me. Employment interviews are more tricky). I did it, too, because I knew there would be a lot of negotiation required to ensure I'd have the supports

Creating a Culture of Accessibility in the Sciences. DOI: http://dx.doi.org/10.1016/B978-0-12-804037-9.00014-0

I needed to work in a lab, and I needed to determine quickly whether my supervisor would be an ally or an opponent in that effort.

During my training, I have been privileged to have some very good supervisors. These individuals have also been my mentors, but I hasten to note that I have had some excellent mentors who have not been involved in any way in a supervision relationship with me. It is really important to draw that distinction – our supervisors have very different expectations placed on them, and, in many ways, there are a lot of requirements in terms of what a supervisor has to do.

During my undergraduate education, I was never particularly close to most of the faculty I interacted with. In a strange way, that made me more open to understanding the importance of the student–supervisor relationship in graduate school. Because I had nothing to compare it to from my undergrad, no real frame of reference, I had no expectations or preconceived notions, and could work with my supervisors to make the relationship what I needed it to be. Very fortunately for me, I had supervisors at all stages of my graduate and postdoctoral training who were very willing to do just that. Those potential supervisors who I could tell weren't as interested, or as open, or as willing to be accommodating of a person with a disability – and there were so many – them, I avoided.

Assiduously.

Say what you will about the motivation of students and postdoctoral fellows in the sciences, and why we choose the training environments we put ourselves in – and how dispassionately we can put ourselves in alien circumstances that are not beneficial to our mental health – but I never personally saw the need to make things more difficult for myself by working for someone who I felt would be unfriendly to the very concept of a person with a disability in their lab. Luck had something to do with my choices of supervisor over the years – but so too did keen observation of the people I interviewed.

I owe a lot to the people who have supervised – and, in so doing, simultaneously mentored – me through my training. Every horror story I have heard, with supervisors being too paternalistic, or supervisors countering accommodation decisions because they felt they knew better, or supervisors and students not having clear expectations of performance and workload, or supervisors reacting negatively to students in crisis, I have managed to avoid. I like to think that happened – or didn't happen, as the case may be – because of a certain amount of hard work on the part of myself and my supervisors. I also think it was because my supervisors were, each in their own way, genuinely invested in the possibility of my success, in a way that ultimately was no different from their investment in the success of their other trainees.

Sure, things were different for me than they were for others in the lab – I had a disability, I required accommodations, and we got pretty creative (particularly during my PhD and postdoctoral training) to make sure that I was able to perform the essential requirements of the science I was doing – but at no point was I made to feel that these things were challenges. Indeed, as my training progressed, I felt instead that I was being an active, valued contributor, who was able to participate in the research group in unique ways all my own.

After all, isn't that the point of being a scientist and working in a team?

Introduction

In the sciences, an important distinction needs be drawn between the *supervision* and *mentorship* roles students will find themselves in throughout their training. In chapter 13: Faculty mentorship of students with disabilities in the sciences, we examined in detail the nature and quality of the mentorship relationship that students with disabilities can have with faculty and their peers. In this chapter, we focus on the supervision relationships that students with disabilities may experience during their training. It is worth noting that, although it is expected that the supervisor will serve as a mentor, there are many situations in which a mentor will not be direct supervisor of the student. There are also certain circumstances where the supervisor is ill-suited for a mentorship role.

Undergraduate and graduate postsecondary education in the sciences involves a series of supervision relationships, through access to summer studentships, internships, co-op placements, fieldwork placements, practicum settings, and the graduate (master's and doctoral) thesis. The student may also be supervised at times by senior students, postdoctoral fellows, technicians, placement supervisors, and faculty members. For students with disabilities, the potential variety and quality of student–supervisor relationships in science, technology, engineering, and mathematics (STEM) poses a challenge in the context of disclosure of accommodation need, and whether disability-related concerns can impact the student–supervisor working relationship.

Later in the chapter, we will return to supervision by senior students, postdoctoral fellows, and technicians, as this *delegated supervision relationship* merits special consideration. For now, we focus on the supervision relationship between a student with a disability and a faculty member.

The foundation of the student–supervisor relationship

At the graduate level, there are formal requirements, independent of discipline, that are outlined by universities and their faculties of graduate studies (or equivalents) for thesis supervisors. Many of these requirements also apply to faculty serving as supervisors of undergraduate students and to placement supervisors in co-op, field-work, and practicum settings. Essentially, supervisors are expected to:

1. Establish clear expectations with the student with respect to communication frequency, research or work program, and milestones for achievement;
2. Provide guidance to students on research or work direction;
3. Monitor student progress toward defined milestones;
4. Provide feedback on accomplishments and progress;
5. Facilitate resources and assistance for the student as necessary based on their progress; and,
6. Evaluate the student's accomplishments at the end of the placement, work term, or program.

The intimacy inherent in a student–supervisor relationship is significantly greater than than that in a student–faculty member relationship, where (unless that faculty member has chosen to mentor the student) the faculty member has considerably less invested in the student and in the outcome of the relationship. These qualitative differences may not be intuitive to students unfamiliar with the supervision setting, but are crucial to the success of the relationship—again, particularly in the graduate research environment.

The potential impact of a student's disability on the nuances of the student–supervisor relationship is significant and has the capacity to influence every aspect of the relationship and every requirement of the supervisor. Because of this potential for far-reaching consequences, students with disabilities need to carefully consider their choices around disclosure of disability and accommodation need.

Disability, disclosure, and the student–supervisor relationship

A supervisor plays a key role in the "socialization" (or integration) of the student in his or her chosen field and career path. The supervisor can serve as a significant resource for professional development and career growth, as well as for the foundation of the student's professional network. Because of the crucial importance of the multiple roles—the academic role, the workplace setting role, and the professional development role, to name a few—to the student–supervisor relationship, *trust* is integral.

This dynamic of trust is influenced by a student's choice to disclose his or her disability and/or accommodation need directly to the supervisor. It is a choice fraught with difficulty from the student's perspective, as there are a number of possible consequences to consider, irrespective of the choice he or she makes. While the student may feel more comfortable navigating the disclosure process through his or her disability services office, imagine a supervisor's reaction when receiving such a letter—without advance notice!—about a student he or she has been working with.

Two scenarios in Chapter 5, Key role of education providers in communication with students and service providers, highlighted some of the choices a student can make when determining whether to disclose to his or her supervisor, and the potential consequences of some of those choices. Simply put, disclosure to a receptive supervisor can be beneficial, while lack of disclosure followed by a decline in student performance or productivity that is disability related can be bad for the relationship and for the student.

It is worth noting that disclosure as an act is only the first step in an ongoing conversation with the supervisor around clarifying expectations, safety, essential requirements, and a host of other relevant issues. The receptivity of the supervisor is as crucial to the success of that dialog as the willingness of the student to engage. On both sides of the conversation, it is important to proceed without assumptions— a student cannot assume the supervisor fully understands everything to do with the functional impact of his or her disability, and the supervisor cannot assume that he or she fully appreciates those functional impacts. There is a fine and very nuanced line to be walked by both parties in the relationship, which adds a layer of complexity to the student–supervisor relationship. The best students and supervisors are able to navigate these nuances collaboratively and effectively.

Quality of the student–supervisor relationship

The student–supervisor relationship is experienced on a continuum from very poor to very effective, with most relationships realized between the two extremes: good in some ways, fair or poor in others. High-quality relationships between students and their thesis advisors (supervisors) are associated with benefits for the university, the supervisor, and the student. These benefits include timely rates to degree completion (Girves & Wemmerus, 1988; Lovitts, 2001), lower rates of attrition (Golde, 2005; Jacks et al., 1983; Lovitts, 2001), and successful socialization into the department and discipline (Gerholm, 1990; Weiss, 1981). Since supervisors often provide career advice, letters of reference to potential employers, and/or further mentoring after graduation, the student–supervisor relationship is one of the most defining relationships of their careers.

In positive student–supervisor relationships, the qualities of an effective advisor include high levels of interaction (accessibility, frequent informal interactions, and connections with many faculty members) (Gerholm, 1990; Girves & Wemmerus, 1988; Weiss, 1981), and purposefully helping students progress in a timely manner (Lovitts, 2001). Students also note flexibility, respect, and strong communication skills as important characteristics of effective supervisors (Skarakis-Doyle & McIntyre, 2008).

Students and their supervisors who have realized high-quality relationships have also achieved a mutual understanding of expectations around the roles and responsibilities of both the student and the supervisor in the student's graduate program. It is necessary to clarify such expectations since considerable variation exists around the roles of both supervisor and student. These roles are negotiated around issues such as funding, graduate student employment, the frequency of meetings, timelines, the type, nature and frequency of feedback provided on written work, authorship and intellectual property, responsibility for thesis topic development and methodology, and the role of other committee members and cosupervisors (Skarakis-Doyle & McIntyre, 2008).

Supervisors may also differ in how ideological or opinion differences are handled and communicated to graduate students (Skarakis-Doyle & McIntyre, 2008), which may be a reflection of their own experience of being supervised.

The variation around roles and responsibilities is likely related to differences in disciplinary cultures and the position of the supervisor. For example, the culture of the discipline or department may determine the format of a dissertation, how the thesis topic(s) are chosen, how the research is conducted, how funds are allocated, and how students and faculty interact (Zhao et al., 2007). In addition, variation among institutions in the requirements of different graduate programs, roles of graduate officers, and policies around the role of the supervisory committee also dictate expectations that must be considered in the student–supervisor relationship.

Deterioration in the student–supervisor relationship

While most graduate students with disabilities have solid and functional—even strong—personal and professional relationships with their supervisors and feel their supervisors are for the most part understanding of disability and accommodation issues, there are some broad examples of student–supervisor relationships that have the potential to break down. According to Sukhai et al. (2016) situations where disability may negatively impact the student–supervisor relationship significantly include the definition and clarification of expectations around productivity and accommodation, student nondisclosure, fear of stigma and the evolution of potential crisis situations, and attitudinal barriers on the part of the supervisor.

In the following sections, we will revisit the issues of expectation, disclosure, and crisis situations. Attitudinal barriers and misconceptions on the part of the supervisor deserve some attention as well. These may take the form of overt paternalism (a "supervisor knows best" mentality) on the part of the faculty member, where the student's participation in key decisions is curtailed; singling the student out personally and professionally within the research group; failing to recognize the bandwidth, mental health, and well-being concerns of the student as he or she works to manage the disability in the context of academic studies; or feeling that the student may be acting in a dishonest manner with respect to his or her disability. In all cases, the (mis) perceptions of the supervisor lead to a level of dysfunction in the student–supervisor relationship, and a deterioration in how well the two individuals work together.

Clarifying expectations in the student–supervisor relationship

The quality of the student–supervisor relationship declines when expectations are not clear to both supervisor and student, or when they are not mutually agreed upon. When the student has a disability, there is a higher risk of mismatched expectations in the student–supervisor relationship as a result of additional factors that must be negotiated by both the student and supervisor. This lack of clarity in expectations can have devastating consequences for the student—the most severe of which includes failure out of the program and removal from the supervisor's research group. The supervisor is also certain to come out of the experience with a negative view toward the participation of students with disabilities in the sciences.

Lack of clarity in expectations can evolve in several settings within a student–supervisor relationship. If the student chooses to work with a faculty member doing research in a field related to his or her disability, the student may assume he or she has greater knowledge about the functional impact of the disability than the supervisor, and as a consequence of not setting expectations for productivity and performance, or to put appropriate accommodations in place. When this happens, the student and supervisor end up working at cross-purposes, as they view things through very different lenses.

The student may also begin his or her work in the research group or lab under different circumstances than ultimately evolve throughout the length of the working relationship. If a summer student transitions into a more part-time role in the lab because of the resumption of classes in the new academic term, e.g., and then struggles with integrating classwork, labwork, and his or her disability, it may appear to the supervisor that the student has lost interest or has productivity issues. Again, in the absence of an appropriate conversation around disability, workload balance, and expectations, the student and his or her supervisor may end up in a dysfunctional relationship, working at cross-purposes.

Many students' disabilities evolve over time and/or may interact differently with the research or lab environment as their project changes. Most accommodations are established upfront at the beginning of a program of study or a placement, and may not take this evolution into account. Thus the student might find themselves in a situation where the nature of his or her research has changed and the functional impact of the environment on his or her disability has also evolved—but, again, without the appropriate conversations with the supervisor, the latter may not be aware of these changes and a dysfunctional relationship can arise. The most egregious example of this type of scenario—students in crisis—merits a separate discussion in the next section.

The power dynamic between student and supervisor, and how it may be altered in the context of disability also bears mentioning. For example, a supervisor may be receptive to other students' challenging his or her views on how to approach a research project, navigate a problem, etc., but perceives the same type of "challenge" from a disabled student as unacceptable. This speaks to the expectation, held by some, that people with disabilities should "do as told." Supervisors then blame these students for not being "a good disabled person" for engaging in dialog in ways considered the norm for their peers.

Students in crisis

The final theme associated with clarifying expectations in the student–supervisor relationship relates to students who have not disclosed their disability and experience crises. Although these situations are rare (some evidence suggests that they happen to less than 1 student in 20), they are the most challenging to resolve in ways sensitive to the concerns and challenges of both the student and the supervisor. Furthermore, crisis situations can result in "last resort" circumstances where the student's autonomy in the lab is lost and reduced. The removal of such autonomy during and after a crisis is humiliating to the student, and the impact on the long-term relationship with the supervisor needs to be taken into consideration.

Crisis circumstances may mean the safety of the student with the disability and/or others is in jeopardy and particular protocols must be followed, resulting in the removal of the student's right to self-disclose to his or her supervisor. As a consequence, the manner in which the student is perceived by the supervisor or others in the department/research group may change, particularly around issues of trust. Supervisors need to be educated about potential warning signs and ways to approach students who may need assistance. To prevent crises from occurring it is important that supervisors provide a respectful and reasonable space for a student to disclose a disability and/or a need for accommodation. It is recommended that supervisors be provided with training on how to respond to students with mental health concerns. The "More Feet on the Ground" campaign led by Brock University in partnership with the Ontario Council of Universities, the Government of Ontario, Niagara College, Pathstone Mental Health, and the Canadian Mental Health Association (Niagara Chapter) was designed to provide free online resources and skill development opportunities to respond to the mental health needs of postsecondary students.

Supervisor's knowledge of and/or willingness to participate in disability-related processes

The supervisor's knowledge of disability and willingness to communicate and/or participate in accommodation processes are important influences in supervisory relationships for graduate students with disabilities requiring academic accommodations to complete their programs. According to research by Sukhai et al.(2016), graduate students who met with their supervisors regarding disability issues did so rarely, and found their supervisors to be receptive of their disability-related needs (including the willingness to provide accommodations).

With so few precedent cases, institutional accommodation policies that aren't specific to students in the sciences, disability services providers (campus student services) who are not equipped with the knowledge and experience to support students with disabilities in the sciences, and the evolving nature of research, some supervisors and university staff will work directly on an ongoing basis with their students to address arising needs for accommodation in a variety of contexts.

In rare cases, supervisors have acted as champions on behalf of students when resistance to accommodation was encountered from other faculty within the department. These situations often arise when the student is not present and in a position to speak for themselves, and can serve as valuable precedents for other faculty, particularly if the advocacy led to successful implementation of accommodations or practices designed to benefit the student.

The role of the supervisor's knowledge of the interface between essential requirements and academic accommodations

Knowledge of disability issues, relevant policies, and issues arising at the interface between accommodations and academic integrity is important for overcoming resistance to academic accommodations and resolving authorship issues. Some students have experienced resistance from faculty, including higher education administrators, around academic accommodations. Among various programs and institutions, there is evidence of an inconsistent application and understanding of essential requirements (see chapter 7: Student as ACTor—recognizing the importance of advocacy, communication, and trailblazing to student success in STEM). It is suggested that faculty be provided with guidelines to assist them in identifying the essential requirements for trainees (i.e., students and postdoctoral fellows) conducting research in their research environment. In addition, the establishment of faculty liaisons equipped with the necessary knowledge to advise faculty on accommodating students with disabilities is also strongly encouraged.

At the interface of research integrity and accommodations: authorship issues

Issues of authorship are another source of conflict arising between graduate students with disabilities and their supervisors at the interface of disability accommodations and essential requirements. For students, authorship is an important element of their research leading to career advancement. The criteria for authorship and associated misattribution of authorship credit has been widely discussed in the literature (Oberlander & Spencer, 2006). A common authorship issue for students is that they sometimes receive less credit than may be expected (Costa & Gatz, 1992). Given this power differential, students may be reluctant to assert themselves when determining authorship credit (Rose & Fischer, 1998). Students with disabilities may be even more vulnerable to authorship issues because of perceived issues of ownership related to the accommodation received for research. For example, a student who is blind may have a laboratory assistant to perform his or her experiments, or a student with a learning disability may have assistance editing a manuscript. Does this mean the person providing the accommodation for the graduate student in research should receive authorship?

To resolve authorship issues around research accommodations, it is imperative to consider the purpose of accommodations. Accommodations are designed to maintain (rather than modify) the academic integrity of the institution (see chapter 7: Student as ACTor—recognizing the importance of advocacy, communication, and trailblazing to student success in STEM). If the most important criteria for primary authorship is how much the author contributed to designing the method, gathering background information, and analyzing and communicating results of the research, then the student with vision loss who has the accommodation of a lab assistant has still met the requirements for authorship as the assistant merely performs the lab experiments. Thus, in dissecting issues related to authorship, the interface between the criteria for authorship and the nature of accommodations must be closely examined. However, it is typically the supervisor who determines the essential requirements of a course, program, or authorship and determines when an accommodation violates academic integrity.

Supervisors may assist students with academic and social integration

Ideally, supervisors for students with disabilities will also assist students in both academic and social integration into the academy. Higher levels of social integration as perceived by students may be one of the most important variables of retention of nontraditional students (Mullen et al., 2003; Tinto 1993). Academic integration refers to participation in the student's institution, while social integration refers to connections with peers and faculty (Mullen et al., 2003). In the research by Sukhai et al. (2016) graduate students who indicated their supervisors were supportive of their disability-related needs noted that their supervisors (who modeled accommodations at meetings with other faculty and students) facilitated their sense of belonging in, and facilitated their motivation to overcome challenges, in their program.

In other cases, students reported their supervisor advocated with administrators so the student could be accommodated at formal thesis defenses (Sukhai et al., 2016). In such situations, the status of a supervisor within the institution (tenure vs. nontenure) may affect his or her ability to initiate support for the student among higher administrators. A supervisor's willingness to assist students with accommodation processes can facilitate the academic integration of students with disabilities into the institution.

In contrast, Sukhai et al. (2016) reported that a minority of graduate students with disabilities indicated a negative relationship with their supervisor. Most students who reported a poor relationship with their supervisor indicated that personality was the primary factor contributing to the breakdown in the relationship and not the disability. A small minority of students attributed a poor relationship with their supervisor to their disability. Some students who expressed explicitly that their disability was negatively affecting the relationship with their supervisor reported that their accommodation needs were not fully understood by supervisors and others in their department. In some cases students felt disrespected when their supervisors implemented their accommodations inconsistently, or continually singled them out asking if they required accommodations. Students with disabilities and their supervisors need to communicate about expectations relating to a student's accommodations, including when the accommodation will be required, how the supervisor will know when a routine accommodation is or is not required, and the processes around the implementation of the accommodation.

Boundary issues

For some students, the good intentions of supervisors to support students with disabilities can result in boundary issues, especially in health care or professional education programs. Boundary issues may arise when the supervisor attempts to take on the role of the disability advisory by assuming he or she knows what is best for the student, especially where the student has not been consulted. In our research, several students reported that their supervisors expressed doubt about the compatibility of particular programs with specific accommodations, even when the accommodations were later granted (Sukhai et al., 2016).

Boundary issues can also arise when a supervisor works in a disability-related space (e.g., a clinician–scientist doing research into degenerative eye conditions, an occupational therapist or occupational scientist who focuses on disability, a computer engineer who focuses on accessible media, etc.), where the student assumes the supervisor understands his or her disability-related concerns. In these situations, the student may assume that discussing disability concerns is not required to facilitate accommodation. When conflict arises between supervisor and student around disability issues, both faculty and students may look to institutional policy for guidance. However, accommodation policies for students with disabilities within postsecondary institutions are often written generally, or with undergraduate students in mind, since that population makes up the bulk of students with disabilities on any given campus. Therefore graduate deans and other administrators in higher education need to develop policies and procedures to address the unique issues faced by students with disabilities in graduate-level programs. It is also important that graduate education guides for supervisors and students include information about how to address conflict around disability-related concerns.

Funding issues

Of particular importance to the student–supervisor relationship is funding around academic meetings. Conferences provide important opportunities for networking and career advancement. Most organizations do not set aside funds to provide accommodations for individuals with disabilities to travel and/or attend conferences. When a student cannot attend a conference, it not only negatively impacts the student, but also the supervisor. For the student, not being able to present his or her work may slow career advancement as many students connect with potential supervisors and/ or employers at conferences. Similarly, for nontenured supervisors, academic meetings present venues for career advancement through the presentation of their students' work and collaboration opportunities, especially for the purposes of securing additional research funding. For supervisors, funding for accommodating students with disabilities in their research groups, when taken from research grants for such purposes, may be viewed as an unfair expense.

Delegated supervision

In many circumstances—particularly with more junior trainees—the faculty member in charge of a research group will delegate the day-to-day supervision responsibility to a senior graduate student or postdoctoral fellow. In situations where technical skillsets need to be passed along to the trainee, a lab technician may be engaged as a supervisor. This poses a challenge for students with disabilities—having chosen whether or not to disclose to the faculty member in charge of the research group, the student must now in effect make the choice again with respect to his or her day-to-day supervisor. However, senior graduate students and postdoctoral fellows are not trained to appropriately manage disclosure. On the other hand, these individuals are often closer in age to the student with a disability who has joined the research group, and under the right conditions, a peer mentorship relationship can form. It is possible that, in those circumstances, the student will choose to disclose to someone they see as a sympathetic peer, instead of an inaccessible supervisor. Indeed, based on the attitudes of the faculty member and the research group, the student with a disability will find him or herself in one of four kinds of situations: An environment where he or she chooses not to disclose to either the faculty or the day-to-day supervisor; a setting where he or she discloses to the faculty member but not the senior student or postdoctoral fellow; a setting where he or she discloses to the senior student or postdoc, but not the faculty member; or, a safe space where he or she feels comfortable enough to disclose to both the faculty member and the day-to-day supervisor.

Conclusion

Since most university systems are not equipped to meet the needs of nontraditional students (Gilardi & Guglielmetti, 2011), students with disabilities and their supervisors sometimes experience conflict around disability issues. The extent of this conflict depends on the attitudes of both the student and the supervisor. In this chapter, we outlined the central importance of the student–supervisor relationship to a student's success in the sciences, some of the key considerations in an effective student–supervisor relationship, as well as some of the pitfalls students with disabilities and their supervisors may encounter in attempting to successfully navigate this relationship.

The student in a leadership, mentorship, and supervision role

Chapter Outline

Introduction   177
Disability management and nontraditional learning environments   178
Employment for students in STEM graduate programs   178
Widening the circle: disclosure when the student is not in a traditional learning
    environment   179
Identifying accommodation needs in nontraditional learning environments   180
Achievement of necessary competencies   182
Stress and nontraditional learning environments   183
Student as trailblazer   183
The student's lived experience with disability—impact on perspectives   184
Conclusion   185

"How can you see to teach?"

That was the question posed to me by the faculty member interviewing me for my first TAship in graduate school. Flush with the heady glow of just having entered graduate school – with the heady glow of having gotten in, when just two months before I'd not been sure of my future direction – I offered what was, perhaps in retrospect, an incautious answer:

"Why don't you hire me and find out?"

I got hired. I spent two years teaching that course.

That was my first exposure to the types of "outside the classroom" or "outside the lab" learning that one gets in the sciences – my first of many, in fact.

I supervised my first summer student within 9 months of starting my Master's. Mentored my first junior graduate student around the same time. Began working with collaborators within the first year of graduate studies. That was a small lab, in a relatively small department, and I don't think my experiences were all that unusual. What certainly made them different, as illustrated by my course coordinator's question during that interview, was my disability and potential need for accommodations.

I had to consider creative ways of teaching students in the lab, given that I couldn't necessarily watch to see them doing things

Creating a Culture of Accessibility in the Sciences. DOI: http://dx.doi.org/10.1016/B978-0-12-804037-9.00015-2

correctly. I had to consider creative ways to teach in a classroom setting. I had to consider whether or not to speak to my students to let them know I needed to do things a bit differently than others. In those days, at the beginning of my graduate training, and not having had any of these experiences during my undergrad, I basically felt my way through the process in order to figure out what was going to work and what wasn't.

When I started my PhD – and then when I did my postdoctoral training – I had the luxury of learning from previous experiences, and so engaging in these different kinds of learning environments became easier. I got more comfortable, I knew what I needed in order to succeed in these roles, and I knew that I could succeed.

I got better at balancing everything, in other words.

I don't use that term – "different kinds of learning environments" – lightly. I very much consider these opportunities for teaching, mentorship, supervision, collaboration and leadership to be professional development and growth opportunities. To me, they are learning of a different kind than goes on in the classroom, in the field or in the lab. They are part of that "hidden curriculum" we referred to earlier. As labour intensive as preparing for them and participating in them might be, I very much valued those experiences as I was having them – value them still today – and considered them to be worth every second of time, every joule of energy that I invested in them. For me, the benefits significantly outweighed the costs.

I've encountered many students – disability or no – who don't feel the same way. Who looked only to the surface learning opportunities, and chose to ignore or engage superficially in these other opportunities. "I don't have the time" is a common refrain. "I don't want to have to be responsible for anyone else's work" is another. "I have too much to manage, and I can't afford to do this too" from some students with disabilities.

Increasingly, though, participation in these additional learning opportunities is required for us to be successful scientists. They are now, more and more, being mandated as part of our training. When I was a student, this was all optional – now, it is expected.

For us, as scientists with disabilities, it means we have to start thinking through the disclosure considerations, we have to think about what we need or not in terms of accommodation, flexibility and support. We need to start thinking about factoring these opportunities into our plans, into our disability management routines, lives and experiences.

Introduction

In a biological sciences research lab, a student takes on a summer student. On an archeology dig, a student is put in charge of a small team of researchers. A student is asked to coordinate an international team of collaborators working on the hunt for extrasolar planets. A student demonstrates techniques in a chemistry lab to her peers. A student leads a computer lab during a tutorial session in quantum mechanics.

In the previous chapters, we addressed issues relevant to the student as a learner, to the student as a mentee, and to the student as a trainee. In the sciences, students are often called upon to take on at least one additional role as part of their training and development—a student may actively teach in the classroom or lab setting; a student may give seminars in formal and informal settings to colleagues, peers, and collaborators; a student may themselves mentor or supervisor their juniors in the lab, fieldwork, or research setting; or a student may take on significant leadership roles within their research groups, or within larger collaborative teams. Sometimes, a student may find himself or herself fulfilling multiple roles simultaneously.

At the graduate and postdoctoral training levels in particular, there is the potential for the student to assume multiple roles in the scientific training setting. Often, discussions around professional development and core competencies of research trainees—particularly in science, technology, engineering, and mathematics (STEM)—make fundamental assumptions about the participation of students in a variety of roles in the lab or research group.

As implemented by multiple professional societies today (see the Core Competencies framework of the National Postdoctoral Association and its related self-assessment checklist; the professional development rubric of Vitae in the United Kingdom; the Federation of American Societies of Experimental Biology's Independent Development Plan for scientific and professional development; and myIDP.org, hosted by the American Association for the Advancement of Science, among others), and as increasingly mandated as part of training and research grants by large national granting agencies (in the United States, see the mentorship guidelines of the National Institutes of Health), a student's or trainee's professional development is evaluated based on their well-roundedness, as defined often by their ability to engage in, willingness to assume, and success in these additional roles.

In order to fully participate in the STEM training environment—indeed, in order to be competitive in increasingly difficult funding climates, students with disabilities need to engage with these additional roles during their training. It is important, then, to discuss how students with disabilities in STEM may fit into these roles and to evaluate the considerations that students, services provider staff, and faculty have to think about and discuss. In particular, what are the considerations around disclosure, the need for accommodation, the achievement and evaluation of necessary competencies, and the potential for stress? How would the student's status as "trailblazer" impact these roles? For the student, how does their lived experience with their disability help (or hinder) their potential for success in these environments?

Disability management and nontraditional learning environments

Students and trainees in STEM may miss out on the types of learning opportunities described in this chapter because of a hyperfocus on succeeding at their own work, or because of a perception (real or imagined) that opportunities such as these would be frowned upon or actively discouraged by their faculty mentors or supervisors. These themes hold true for students and trainees with disabilities in STEM fields as well—with an added complication: a student with a disability often expends significant effort "managing" their disability in daily life, which can be both energy- and time-intensive. Students with disabilities are often thought to require more time to complete programs of study (although the literature supporting this perspective is somewhat limited and unclear), at least in part because of disability management, in addition to potentially needing more time to prepare for and complete tasks.

Disability management is often used as an argument for why students with disabilities may choose to not engage in co- or extracurricular activities while in postsecondary education. Since the business of living and managing one's disability(ies) needs to be factored into daily life and daily planning, doing "more" outside the basic requirements of school and employment is considered a luxury. Coupled with the potential for inaccessibility of and lack of accommodation for anything not considered part of the "core" academic curriculum, students with disabilities often choose not to engage in those types of activities.

In the context of STEM, specifically, and research training more generally, the student as a mentor, supervisor, instructor, collaborator, and/or leader forms a significant fraction of a "hidden curriculum" in education and training, particularly as increasingly defined through the competency frameworks identified above. These activities are not co- or extracurricular per se—and particularly in the graduate and postdoctoral settings, are considered to be an important part of the training environment.

As the student may not always be aware of this perspective, open conversations about engaging in these additional learning and growth opportunities should be had. Recognizing that such learning and growth opportunities as we have discussed so far cannot always be planned for in advance is also important as well, as a student armed with this information can take it into account in determining his or her own individual disability management approaches.

Employment for students in STEM graduate programs

In addition to classroom, research, and laboratory work, students at the graduate level in STEM programs may be expected to take on employment responsibilities during their studies. In fact, this is quite common in any research-based graduate-level program. Employment may take the form of teaching assistantships, research assistantships, graduate assistant work, or laboratory assistantships. One common challenge for students with disabilities who navigate these academic employment settings is implementing needed accommodations in the employment context (Sukhai et al., 2016). Disability services offices are responsible for providing in-the-classroom accommodations, and strive to ensure students are accommodated within their program of study. However, many teaching and research assistantships are considered paid employment, and are not therefore directly linked to classroom learning. This is where accommodation outside the classroom becomes an important conversation, and one that needs to take graduate employment into account. Students will need to evaluate the need for disclosure of their accommodation requirements (see chapter 6: Disclosure in the sciences) separately from their academic program. Students may be fortunate enough to work for a supervisor (who may or may not be their thesis supervisor) as a teaching or research assistant who is flexible and willing to think creatively about the essential requirements of the job duties (see chapter 10: Essential requirements and academic accommodations in the sciences). Challenges with accessing accommodations include obtaining needed disability-related funding to perform the duties of the academic employment, acquiring materials in an accessible format, and negotiating extra time to complete tasks. Given the complex interface between academic employment and accommodation, it is necessary for institutions to develop policies and procedures for accommodating students with disabilities in academic employment and ensure this information is clearly communicated to students, unions, faculty, and departments (Sukhai et al., 2016). It is essential to have a designated pool of funds for accommodating students with disabilities in academic employment settings, along with a clear process outlined in order to access this funding, and ensure students, unions, faculty, and departments know how to access this funding. To ensure students, faculty, and teaching assistant unions have the necessary tools and knowledge to implement accommodations in the graduate employment environment, resource guides for students with disabilities and student leaders (i.e., teaching assistant union and students' association leaders) need to be developed around barriers to, and accessing accommodations to, academic employment.

Widening the circle: disclosure when the student is not in a traditional learning environment

Imagine: You are a student with a disability, and you are asked to lead a seminar on your work for a room full of departmental colleagues and collaborators. This informal setting can be very stressful, particularly if you require accommodations to prepare for and/or deliver the seminar. Who do you talk to about setting up the appropriate accommodations? Particularly if you are given short notice, and are unable to secure a meeting with your disability services office in time? Refusing or rescheduling this opportunity may not be possible, or may reflect poorly on you.

Or imagine: You are asked to lead a collaborative project team for a period of time. While your participation on the team was manageable with the accommodation set you had previously negotiated, a new accommodation plan is required if you are leading the group—and you may end up inadvertently disclosing your disability due to the accommodation, even if this were not your intent.

The student may have chosen to disclose his or her disability to the campus disability services office (see chapter 6: Disclosure in the sciences), but involvement in these nontraditional learning environments may put the student in circumstances where the disability services staff are unable to assist at all (e.g., if the learning environment falls outside the scope of their operations) or unable to assist in a timely manner. Because disclosure of an accommodation need is a personal choice, it is important for the student to fully consider the potential ramifications of disclosing—or not disclosing—to additional individuals outside the circle he or she has already chosen to establish.

Many of the possible paths available to students depend on whether they have chosen to disclose to faculty members or supervisors, and the reception their disclosure received. If it was positive and encouraging, the student may have an ally or champion in navigating the possibilities around disclosure in these new environments. On the other hand, if the student's disclosure was received negatively by the faculty or supervisor—or the student had chosen to not disclose—he or she may have to choose whether or not to identify a potential ally or champion in the environment or circumstance he or she is in, and consider disclosing to that individual in order to solicit the appropriate assistance.

The following are some of the questions the student may consider when approaching a nontraditional learning environment:

- What is the nature of the environment? Does it constitute collaborate group work, for example? Seminar presentations? One-on-one mentorship or supervision? Something else?
- Are interactions in this environment primarily with peers or colleagues? Faculty? More junior trainees? What is the nature of any potential power dynamic?
- Is the student in a more visible leadership role?
- What potential tasks may require accommodation, which will in turn influence the choice to disclose?
- In interactive settings (e.g., mentorship and supervision of a trainee), is it necessary for the person(s) the student interacts with to know about the accommodation need?

Specifically, students must consider *who* they may need to disclose to, *what* they specifically need to disclose, and *how* to facilitate that conversation. While students must parse the potential impact of expanding the circle of knowledge of their accommodation need against their potential for professional development and career advancement, they must also do so against an honest evaluation of whether the disability services staff or faculty they have disclosed to previously are in a position to be of assistance.

Identifying accommodation needs in nontraditional learning environments

Understanding the functional impact of a student's or trainee's disability in the teaching setting—classroom or lab—or as a mentor or supervisor or in collaborative group and leadership settings is important in determining his or her appropriate accommodation needs. It is easy for everyone involved (student, services provider staff, and faculty) to assume that accommodation may not be necessary outside the traditional classroom or laboratory or fieldwork environment; however, this may be incorrect once a task breakdown exercise is undertaken, akin to those outlined for the laboratory setting in chapter 17: Accommodating students with disabilities in science laboratories and in fieldwork.

Consider that the functional impact of the instructor's disability in the traditional classroom setting is well-understood: The potential need for additional preparation time, utilization of alternative teaching styles as appropriate, and flexibility in designing and implementing evaluation and assessment methods. However, translated to a laboratory environment, the functional impact of a disability in an instructional context is both subtly and overtly different. For example, what assistance or accommodation does the student need in order to effectively demonstrate a lab technique to the class? Or to monitor the effectiveness of students' laboratory approaches? What additional preparation may the instructor need in order to function well in the teaching lab? And how can teaching styles and assessment approaches be adapted as necessary in response to the instructor's accommodation needs?

A similar thought process applies to situations where the trainee with a disability is in a mentorship, supervisory, collaborative, or leadership role. Accommodating the student in the performance of his or her own laboratory or fieldwork tasks is one thing, but those accommodations often won't translate to circumstances and scenarios where the student becomes the teacher—not only in a group setting, as highlighted above, but also one-on-one, as is more likely in a mentorship or supervisory relationship in the lab or in the field. A student with a disability may have evolved appropriate and effective adaptations in the lab or in the field to manage the functional impact of his or her disability, and would be able to do so again with preparation and practice when asked to take on an instructional role. For example, a student who is visually impaired and engages a technical assistant in performing his or her own laboratory experiments may ask the trainee to verbally describe every step of the experiment he or she is running (both the method and the results or outcomes), and then talk the trainer through how to troubleshoot any errors, in much the same way he or she would work with a technical assistant. Or a student with a learning disability may design his or her work with a trainee as a "learning exercise" with defined expectations, goals, and outcomes. Or a student who requires additional time to process questions may ask his or her trainee to email questions, as opposed to asking and answering questions during class.

Ultimately, regardless of the setting, a common exercise or thought process needs to be worked through by the student, as well as (wherever possible) the faculty member/supervisor and the disability services staff and should include the following questions:

- What is the nature of the specific task(s) to be performed by the student? Does it involve group instruction? Individual instruction? Group collaboration? One-on-one discussion? Presentations? Other activities?
- What are the specific essential requirements of these tasks and learning outcomes for the student? Are they modifiable based on the student's needs?
- Are there any barriers to ensuring the student's full participation in these activities?

- Does the student require any assistance or accommodation to perform the tasks in order to achieve the required competencies?
- Will this assistance or accommodation interfere in any way with the defined competencies?
- Are there ways to adapt the student's current classroom, laboratory, or fieldwork accommodations in these additional contexts?
- Are there aids, tools, and technologies (mainstream or assistive) that may be beneficial to the student in these settings?

It is crucial to recognize at this stage that due diligence by the student, faculty, and services provider around disclosure and accommodation in settings where the student does not assume traditional "learner" roles is not intended to lead to an outcome where the student is denied the opportunity to participate. For students with disabilities in STEM to have equal opportunities in career and professional growth, being able to engage as a mentor or supervisor, being able to teach in group settings in the lab or in the field, being able to present informally and formally in seminars and journal clubs, and being able to participate collaboratively and in a leadership role in larges projects are important to their growth and development.

Note that we are not arguing that circumstances must be forced, through accommodation or flexibility of essential requirements, into an alignment to guarantee student success—indeed, students in STEM generally are not guaranteed to succeed in such settings, at least some of which has to do with personal attitude and recognition of these opportunities as learning or professional development avenues. Instead, recognition that students with disabilities may be excluded from, or exclude themselves from, these opportunities is important—as is an understanding that students may not know these opportunities are possible, or choose to seek them out. Thus the conversation around accommodation is at its core a conversation around facilitating equal opportunity for student engagement.

Achievement of necessary competencies

How does a student recognize his or her growth and development through engaging in leadership, mentorship, supervision, and collaboration opportunities as part of the STEM training? In some circumstances, students may be able to receive feedback on their approaches and execution. Certainly, if they make a mistake that warrants comment, such feedback is provided. Many students will pick up on their achievements and successes.

If students put forth the effort required to participate in the kinds of nontraditional learning described in this chapter, and to understand and appreciate how to manage their disabilities in these contexts, providing them objective feedback on their success is important. Indeed, for some students—e.g., those on the autism spectrum—this objective feedback provides structure and definition to the activities they are asked to engage in.

For faculty, services providers, and students seeking some framework around engagement in the nontraditional learning environments we discuss here, independent development plan (IDP) frameworks—such as those developed by the Federation of American Societies of Experimental Biology—can prove beneficial. These models allow for student self-assessment, objective feedback from faculty mentors, and identification in a rigorous manner of learning and professional development activities that would benefit students. Furthermore, with slight modification, IDPs can be adapted to include appropriate and sensitive conversations around disability and accommodation need, as well as both student and faculty concerns around career and professional development, in the context of disability management.

Stress and nontraditional learning environments

Involvement in nontraditional learning roles can be stressful for students. The added responsibility coupled with the need to find time to coordinate new roles with their existing workloads as students can be daunting. While some students revel in the challenge, others shy away from the added work and increased responsibility. Students with disabilities are no exception—indeed, the potential for stress is elevated, as students may be concerned about how to effectively manage their disabilities given the added work, about their potential accommodation needs and the logistics of their implementation, or about potentially disclosing to a larger circle than they had originally planned for. New roles, responsibilities, and learning environments also increase the chances of experiencing impostor syndrome, and the time taken to prepare for and work within these new roles may, at least initially, reduce the time required for self-care (see chapter 8: Mental health and well-being for students with disabilities in the sciences).

Recognizing the potential sources of stress for students as they integrate into additional roles and learning environments in their training can help faculty and services providers work to identify appropriate coping strategies and potential accommodations. For students with disabilities, acknowledging potential sources of stress can help in preemptively developing appropriate coping and accommodation plans and can mitigate any issues that the transition into an added role and learning environment causes.

Student as trailblazer

With any new role comes the possibility of the student becoming a trailblazer in his or her chosen STEM field. A student may not be the first person with a disability in archeology at his or her university or college, but may become the first to lead a collaborative team at a dig site. Or, he or she may not be the first person with a disability in chemistry, but may become the first to serve as a teaching lab demonstrator.

With each foray into new territory, there will be individuals—peers, junior students, faculty members, collaborators—who may harbor doubts and may further choose to express them vocally. Thus the student is forced to confront this trailblazer status over and over again, and sometimes be put in a position to have to defend his or her capabilities. In these scenarios, having a faculty champion—mentor, instructor, supervisor—who is willing to counter any concerns is greatly beneficial and can potentially reduce the student's stress. It is important that the student is aware that these issues and conversations may arise, and that a champion is speaking on his or her behalf; it is also important for the faculty champion to be sensitive to the potential impact on the student's well-being in speaking to him or her about this issue. The timing and setting of these conversations can be important in helping the student to manage the stress of being a trailblazer.

Perhaps harder to address because of the power dynamic between the student with a disability and these individuals are the doubts harbored by peers, colleagues, and/or junior students. Particularly if the student has chosen to disclose his or her accommodation need, he or she may encounter bias and discrimination around engagement in the role as an instructor, supervisor, mentor, or collaborator. It may be difficult for a faculty champion or mentor to actively step in on the student's behalf, because this may set the wrong precedent with respect to the student's relationships with his or her trainees. In a teaching situation, where the student with a disability is responsible for his or her trainees, it may be simpler to acknowledge that someone may not want to be taught by that student and to simply let them move on to someone else. If the trust required for a positive teaching and learning experience cannot be established, a parting of ways is sometimes best.

The student's lived experience with disability—impact on perspectives

As discussed earlier in this book, a student's experience with living with a disability, comfort level and identity, and willingness to disclose accommodation needs will significantly impact his or her experience as a learner, as well his or her insistence on self-reliance. Students for whom their disability was recently identified or manifested, and students who had negative early experiences around their disability and accommodation needs, will have significantly different perspectives, learning, and advocacy approaches than will students with more positive early experiences. This is true in the context of both the formal and informal roles the student may play as leaders, mentors, teachers, supervisors, and collaborators in his or her professional growth as well—indeed, in many ways, even more so, given the relative informality of many of these roles. The student's lived experience and perspective may manifest itself overtly (through level and type of preparation, request—or not—for accommodation, and direct personal interactions) or more subtly. Recognition that the student's attitudes may be due to his or her lived experience can be important when working with the student in leveraging the opportunities presented by nontraditional learning environments.

Conclusion

In the sciences specifically, and the research training environment more generally, students are often called on to assume different roles than the traditional learner, mentee, or trainee roles. In these circumstances, students may themselves be asked to model leadership, mentorship, supervision, or collaboration approaches to their peers, colleagues, and junior students. Faculty members working with students with disabilities in STEM should recognize that a number of factors can influence a student's choice to engage with these opportunities, as well as his or her relative success in that engagement. In this chapter, we offered several insights into the potential barriers for students with disabilities, and the considerations for faculty, services providers, and students with disabilities to facilitate effective engagement with and success in these nontraditional learning environments.

Leveraging professional development and networking opportunities

Chapter Outline

Introduction   190
Types of networking opportunities   190
Disability-specific structured professional networking activities   191
Networking through conferences, career events, and symposia   192
Peer networks and collaborations   192
Informational interviews   193
Creating your own portfolio of networking opportunities   193
Framing disability in networking   194
Defining your personal story or brand and its impact on networking   194
Receptivity of your network   195
Conclusion   196

Amanda's Story: During the summer of 2014 I had the privilege of attending a conference in Seattle, Washington for legally blind professionals working in science, technology, engineering, and mathematics. It was here where I met, for the first time, other "blind" professionals with a science or mathematics background. I was one of 3 biologists at this event and it was also where I met another legally blind geneticist. Dr. Mahadeo Sukhai is a PhD holder from Canada and even though his area of study is cancer genomics, he and I enjoyed many stimulating conversations together regarding sequencing technology, adaptive laboratory equipment, and teaching biology at the college level.

By the final day of the conference, Dr. Sukhai and I became comfortable discussing our experiences as scientists with visual disabilities. I will always be grateful to Dr. Sukhai for being honest with me about the difficulties he experienced as a PhD student and as a postdoctorate, but the message was simple. My goals and aspirations were entirely obtainable. Since this conference, I have been fortunate to have Dr. Sukhai to consult with and his guidance has been invaluable to me. When I reach the point in my life when I am in a place to offer advice to the next generation of prospective graduate students (whether they are people with or

Creating a Culture of Accessibility in the Sciences. DOI: http://dx.doi.org/10.1016/B978-0-12-804037-9.00016-4

without disabilities), it is this type of mentorship which I most aspire to be able to provide.

Amy's Story: I find it helpful to frame my disability as a chance to have unique conversations with others and create an interesting frame around why I do the work that I do or why I am passionate about it—I use it as a means of facilitating dialog and connections. I try to go to networking events and sessions less focused on my experience with disability and more open to learning about others and then taking that perspective home with me to consider in terms of my own growth and development. I only really attend networking events that are specific to my industry or job function – where it makes sense to or adds to the conversation I discuss disability – but the decision to bring up conversations about disability is more a matter of judging in the moment if that needs to be at the forefront or if it is relevant to the conversation or relationship building.

I also think that being able to communicate and collaborate directly with others with disabilities in my network helps to reframe the whole dialog around advocacy, inclusion, accessibility, etc., because you quickly move away from your own silo'd experience and come to understand these issues from a broader and more objective perspective that better serves a greater number of people.

I seek out scholarships, fellowships, networking events, industry events or Q&A sessions, panels, etc., to take a comprehensive approach to building my network. I try to mix up my approach so that I meet different people and learn different things – sometimes I am more functional-role focused and other times I am focused on building my more personal advocacy-focused network.

I seek out people of all ages to build my network—knowing that younger generations can offer me great insight into how I can help advocate for younger people as well as my own peers. I think mentorship is closely linked with any kind of networking of this variety because you end up meeting younger people who become mentees who want to benefit from your experience with disability and who you want to impart knowledge and insight to – and you also have your own mentors and those who are paving the way in the disability space and who you see as leaders.

I always try to think of networking as how I as a person can contribute to a community of people or to an area of thought leadership – my disability may help me frame how I make the contribution, but I see it as a factor that comprises a diverse perspective more than a driving factor.

Disability and vision loss are a part of my story – they have contributed to many of the traits I value most that I would say are part of my personal brand – resilience, determination, a relentless work ethic, creativity, effective listening, empathy, courage, and confidence are all traits that have come to be a part of the brand I communicate to the world – and they all relate to my experience as a person who is legally blind.

Knowing how disability has contributed to who I am and my outlook on the world is important for networking because it allows me to build effective relationships with others.

Mahadeo's Story: My network is the only reason I'm currently employed. My postdoctoral fellowships were secured through my existing network and the reputation I built for myself during my doctoral training. My current position was developed in much the same way, a few years later.

They say it's not what you know, but who you know – let's couple that with the significant and systemic challenges in front of a person with a disability when seeking employment generally, let alone in the sciences, where every attitudinal barrier that applies at the education level is magnified in the employment environment, without the support and protection of accommodation specialists and institutional obligations.

I was fortunately in the position where I knew a few people.

I assembled my network, one person at a time, through a more basic approach in the early days: I did it by building my reputation. I did it by, day after day, demonstrating competency and capacity. Actions speak louder than words? I spent the entirety of my doctoral studies, in fact the entirety of my scientific career, modeling that truism.

Today, I take advantage of conferences and event-based networking, as well as of leadership and fellowship opportunities, as Amy describes – it took me a while to develop the courage to do that, though, because I'm a strong believer in letting my accomplishments speak for themselves.

The problem with that approach is that they don't, in fact, speak for themselves. We are almost expected to talk them up, and if we try to be unassuming about it, no one listens.

Today, too, I take advantage of the fact that somewhere along the line, I went from being the information sponge in a networking event, to being an information provider. Please don't misunderstand– I'm still a sponge – it's important to always network with the intent to learn. But, now I am also in a position to give back from a content and knowledgebase experience perspective, over and above that which my own lived experience in the sciences offers. For me this is a powerful resource, because it eases my comfort level with the engagement – I feel like I am offering meaningful contributions on several levels. Amanda's story illustrates that people I meet in networking events feel the same way, and that my contributions are real and measurable.

Introduction

Student professional development is an increasingly important part of the postsecondary education framework. Many schools in North America have established integrated professional development opportunities into curricula at the undergraduate and graduate levels, and a exposure to and development of the "soft skills" set is a key component of practicum and coop placement programs (for a discussion of the consideration of accessibility in the practicum setting, see chapter 24: Practicum placements).

In the sciences, professional development may take a variety of forms, both formal (e.g., professional development curricula, conferences, and symposia) and informal (e.g., collaborative opportunities, as well as informal mentorship, teaching, and leadership roles; see chapter 15: The student in a leadership, mentorship, and supervision role) in nature. The recognition of these opportunities and the willingness to take advantage of them by the student is very important and increasingly relevant to student success.

Inextricably linked to professional development opportunities is the concept of networking. Networking in the sciences differs little from networking in other professional areas. However, networking in the sciences for students with disabilities is nuanced in ways requiring some discussion. Therefore in this chapter we focus on the importance of networking in a professional development context, via both formal and informal opportunities, and offer strategies for success that a student with a disability may apply.

Types of networking opportunities

As with peer support (see chapter 9: Peer-support networks), networking opportunities may come in a variety of forms, both formal and informal, in group settings or through individual interactions, through external facilitation or self-organization, and may occur through in-person interaction or virtually or online. The type and quality of networking may also be defined by what type(s) of interactions you, the student, are looking for—are you more interested in the experiences of fellow students with disabilities in your discipline, of professionals in your field, in the sciences generally, or in higher education? Or, are you interested in academic and employment experiences in your field of interest, irrespective of disability?

Your school, faculty, or department may have a number of networking opportunities already created, in the form of formal and student-led career development programming and special-interest groups. Formal and informal networking opportunities may be created locally, in your department or campus, by students or by career services professionals on campus, or through faculty efforts. Looking beyond your individual institution, you may find special-interest groups, committees, or trainee councils (with a specific focus on career development and networking) created by professional societies in your field or discipline. In the diversity space, there are several professional societies and nongovernmental organizations that work in STEM and diversity broadly—a few of which may have specific interests in disability, high school, and/or postsecondary education. Additionally, you may find, through a variety of nongovernmental organizations, disability-specific or cross-disability youth councils or professional development programs that are oriented at least in part toward fostering professional networks for youth with disabilities (although these are not necessarily oriented toward educational contexts, and would definitely not be focused on any specific field of study, let alone anything in STEM). Several grant-funded projects have been created to provide formal networking opportunities to students in STEM, but these have not been sustainable in an international or truly connective sense in the long term.

Thus networking opportunities suiting your specific needs may be left entirely up to you, as a student with a disability, to create or find. Faculty, educators, and accommodation specialists can be helpful if they are aware of resources (e.g., locally, online, or internationally) beneficial to students in terms of identifying a professional network.

Disability-specific structured professional networking activities

We will spend some time on disability-specific structured professional networking activities, as these are a unique synthesis of disability-associated networking and networking in the context of STEM and career development. An example of such an activity would be a workshop that brings together students studying in the sciences with mentors with disabilities who work in STEM. This may happen at a local level, on a small scale, or could happen on a national, or even international, level, if organized by groups with sufficient resources.

Structured networking is most useful when you (the student) choose to go with an open mind and engage with all other people in the room or at the session; you also have to be mindful of not just "what am I hoping to get out of the experience" but also "what can I offer others in the room." Indeed, being able to articulate this balance is important as it will foster a spirit of open-mindedness and participation.

Different types of outcomes can arise from these structured professional networking activities—career mentorship, meeting people, communication improvement, and professional development are among the possible outcomes.

Career mentorship may arise through dialog with mentors who are present, who are willing and able to share their own academic and work experiences, and who can offer guidance with respect to your own career trajectory.

Meeting people is possible through being open to interacting with others at the event, and recognizing the potential to grow both a peer-support framework (see chapter 9: Peer-support networks) as well as a professional network that includes mentors from outside your typical mentorship circle (see chapter 13: Faculty mentorship of students with disabilities in the sciences).

Communication improvement occurs because you must articulate your career goals, as well as your goals for the workshop you're attending, to your peers and mentors. This process empowers your development of an "elevator speech" about yourself, your interests, and your career path, and helps define reasons for why professional mentors should invest effort in your development. Furthermore, it is also good to get out of one's own comfort zone for networking conversations.

Professional development happens because networking events of this nature are often designed around some measure of skill learning and growth of the attendees.

Structured networking is most successful when you demonstrate flexibility in resolving and managing your expectations for the workshop or event—it is important to be responsive and adaptive to the people in the room with you. If you don't adapt, you will not be able to derive value from your participation in the workshop.

Disability-specific networking events in the sciences provide a safe space to talk about disability in the sciences. This is very distinct from career networking events independent of disability, where having discussions around disability becomes impossible without being willing to disclose to strangers in an unfamiliar setting (see chapter 6: Disclosure in the sciences). Safety around disclosure is also only possible if both the students and mentors are persons with disabilities; in a setting where the mentors are not individuals with disabilities, but the students are, disclosure is again fraught with challenge—in fact, in those settings, the choice to disclose has been effectively taken away from the student by the nature of the event.

Networking through conferences, career events, and symposia

Conferences, symposia, and career events (particularly those specific to your program or discipline) provide opportunities to network with professionals in your field in settings where your primary goal is career mentorship and expansion of your professional network. There is significant value to face-to-face relationship building through career-related and/or academic conferences and groups, as this avenue provides you an opportunity to make a positive impression through dialog and discussion of your work. Since conferences and symposia also often provide opportunities for you to present your research (particularly true for trainees at the graduate and postdoctoral levels), the presentation and discussion of your work often becomes a "conversational hook" upon which to build the networking dialog. In this setting, it is crucial to approach people.

In career events, symposia, and conferences, it is important to go with an "elevator speech" already prepared—why are you here, what are you hoping to gain, who are you hoping to meet?

Peer networks and collaborations

"Team science" is increasingly important in the research setting today. In the sciences, students are encouraged to build their peer networks at an early stage—even as early as their undergraduate education. Peer networks can be built around shared experiences and potential for collaboration, or they can be built around disability (or both; see chapter 9: Peer-support networks; for a discussion of the importance of peer networks in a support context, and chapter 15: The student in a leadership, mentorship, and supervision role; for a commentary on the importance of thinking through disability, disclosure, and accommodation in the collaboration setting). Collaborations arising from peer networks that are welcoming of a research perspective informed by your lived experience with disability embody the spirit of diversity of thought in the sciences, and are enriched by the dialog that results. Provided that benefits to the research results are evident from this process, the result is a validation of the inclusion of the lived experience perspective, and a demonstration of the value of a dialog inclusive of disability and accessibility. A concrete example of the preceding might be the impact of disability and accessibility on study design and research method, which may in turn lead to a more wholesome study than may have been developed otherwise.

Even when a peer network does not have other students with disabilities, one's disability and lived experience can be channeled into an environment of peer fellows working toward improvement in diversity and inclusion awareness, as well as progress in both academic and professional settings. You are then able to contribute in a collaborative team environment because you can move beyond advocacy into taking action with a peer-support network and mentorship; you have also been chosen to represent the cause and therefore gain legitimacy in what you are seeking to do as a leader in that specific area.

Informational interviews

In the context of disability in STEM, informational interviews as a networking strategy can take several different forms: They are ways to understand and gather information about the next steps in training and career development, ways to identify potential mentors or employers, or ways to discuss disability in the context of the sciences. Ultimately, an informational interview functions to equip yourself with a behind-the-scenes look at a particular topic. These different purposes will lead to different discussion points during an informational interview, and require the student to be prepared for multiple avenues of conversation in a short timeframe. Because informational interviews are often time restricted in nature, it is important to make sure you, as the student, don't oversell or have the other party misinterpret your conversation during the informational interview. A key point to remember is that, although you are in control of the discussion points and agenda for the informational interview, you are being evaluated by the interviewee, as much as you yourself may be doing the interviewing.

Creating your own portfolio of networking opportunities

A reasonable expectation of students in the sciences is that they should evolve their own collaborative, mentorship, and professional development networks over the lifetime of their undergraduate, graduate, and postdoctoral training. As discussed in this chapter, disability adds a layer of complexity to this networking thought process, but does not detract from its value or the responsibility of the student.

To build your own network, it is important to think through a "networking strategy": Where do you go to seek out networking and collaboration opportunities? What mix of scholarships, fellowships, networking events, industry events, conferences or Q&A sessions, panels, etc., are most appropriate for you, given your field? What balance of collaboration, mentorship, and professional/career development are you seeking? What balance of functional-role and personal advocacy-focused networking are you interested in? How do you frame disability in the context of networking (see the subsequent section)? If you are interested in informational networking or informational interviews, do you have individuals or organizations in mind to approach? And if not, how do you go about identifying these?

Framing disability in networking

It is important for the student to consider to what extent he or she may wish to disclose in the context of networking, while mindful of the structure of the event, and whether that structure has removed the choice of disclosure from consideration. As highlighted in chapter 6, Disclosure in the sciences, students have the right to disclose, and students can also be trailblazers (i.e., advocate, communicator, and trailblazer (ACTor)) depending on how comfortable they are with disclosure (chapter 8: Mental health and well-being for students with disabilities in the sciences). We follow up on that thread here, and note that one's disability, framed effectively in the networking conversation, can be a powerful way to have unique conversations with others, and to create an interesting frame around why you do the work you do or why you are passionate about it. In other words, one's disability is used as a means of facilitating dialog and connections. In different formal networking opportunities, it is a chance to connect the dots between career path and lived experience. It's important for people with disabilities to feel open to building networks and knowing that they have a unique perspective to contribute to these types of events; disability also shouldn't deter someone from attending such an event—people are very open and willing to learn and interact in these environments if you give them a chance and if you yourself feel your voice is worth being heard.

Defining your personal story or brand and its impact on networking

Knowing yourself and how to communicate your own outlook and perspective on disability is critical to affecting change in the world around you—at school, work, in a career field, etc. An authentic approach to personal brand where you acknowledge not just what you're good at, but what you've struggled with and how it has benefitted you or positively changed you, is just as important to communicate. In networking and in life you want to have a clear and authentic personal brand—building traits you have gained or perspectives you have gained through disability into that brand message helps you tell a unique and memorable story—because it sets you apart.

If your work is in a related field to your disability (e.g., health care) then you also gain brand legitimacy in that you represent the community you endeavor to serve; this can be very helpful when building your network because the "why" around what you do and what industry you work in comes naturally. You have a personal connection to talk about at an event or over coffee or when meeting a younger person or an older mentor but you can also share how you carved out your professional career path and the challenges involved therein.

Receptivity of your network

Incorporating disability into the networking conversation requires an assessment on the student's part of the receptivity to that dialog of the professional or peer. The onus is on the student to determine how to best frame his or her story in response to the person he or she is talking to. Much of this is dependent on the student's disability (visible vs. invisible), his or her willingness to engage in conversations around disability and accessibility (see chapter 6: Disclosure in the sciences), the impact of that willingness on his or her engagement as an advocate, communicator, and trailblazer (see chapter 8: Mental health and well-being for students with disabilities in the sciences), as well as his or her perspective and lived experience (see chapter 4: Student perspectives on disability—impact on education, career path and accommodation). A student's effective engagement with his or her network is also somewhat dependent on his or her well-being and mental health (see chapter 7: Student as ACTor—recognizing the importance of advocacy, communication, and trailblazing to student success in STEM).

Networks—especially those in a professional setting—respond well to demonstration of competence. First impressions are crucial—it is up to the student to determine how best to drive that conversation effectively. By analogy with how mentors (see chapter 13: Faculty mentorship of students with disabilities in the sciences) and education providers (see chapter 5: Key role of education providers in communication with students and service providers) may respond to disability, members of a professional network may respond to disability—a disclosure or a conversation—in very different ways. While some network members may be receptive to dialog, others may be dismissive either of the impact or relevance of disability on lived experience, or of the student's own competency as a result of disability. Network members may be tempted to play the "gatekeeper" role as well (see chapter 3: Barriers faced by students with disabilities in science laboratory and practical space settings and chapter 10: Essential requirements and academic accommodations in the sciences). Strategies we have previously proposed to address some of those scenarios are also applicable here, with the nuance that a networking relationship is entirely voluntary, both on the part of the student and the part of the professional. While a student with a disability cannot force an interaction where one is rejected, the student also has the choice to abandon a relationship if it is clear that the professional is unable or unwilling to provide the level of interaction and information the student seeks.

Conclusion

Networking provides an important opportunity for continual learning, growth, and professional development and should be taken advantage of as part of a student's journey through STEM education. Indeed, networking should be fostered, where possible, as early as can be arranged in a student's educational journey. The right network and the right mentorship can help guard against some of the attitudinal barriers identified in the education system (see chapter 3: Barriers faced by students with disabilities in science laboratory and practical space settings).

It is especially important to maintain the continual learning frame of mind with respect to networking as a person with a disability, as this strategy may also help to level the playing field for yourself in STEM and for others down the road. A collaborative network will serve as the bridge between academic and social networking. Social capital and the development of your reputation are crucial elements of both your social network and networking as a whole, and provide tangible measures of benefit back to you, the student, in addition to the information and knowledge gained through networking. Leadership in your networks as a means to build your reputation and social capital is also related to this concept. As we have touched on in previous chapters, you can play a peer leadership role as a student with a disability by thinking systemically, which is often of benefit to you as well as to your network.

Part VI

Accommodating Students With Disabilities in the Sciences

Accommodating students with disabilities in science laboratories and in fieldwork

Chapter Outline

Introduction   200
Teaching practices, supports, and accommodations   201
Activities in a teaching lab setting in the sciences   202
Accommodation in the graduate research laboratory   202
Accommodation in the fieldwork setting   203
Discussing accommodations in the science lab or fieldwork settings   204
Conclusion   205

After nearly twenty years in a research lab setting, and four years before that in various undergraduate teaching labs, I've gotten very good at quickly figuring out my lab accommodations. I've benefited from faculty and staff around me being tremendously creative – one of the advantages of being first in my field, I suppose, is that once people got over the shock of having a blind person in their lab space, there was a recognition that there were no rules – and, hence, no limits.

I've gotten good over time at encouraging people to have the right conversation around integrating a person with a disability into the science lab environment – there are so many things to take into account! What are the tasks to be performed? What kind of research or teaching setting is it? What discipline are we talking about? Will the student work in groups? If it's a teaching lab, how are students being evaluated? What's the functional impact of the student's disability(ies)? What are the safety considerations? What are the physical or technology challenges? Does the student benefit from a laboratory assistant? These questions are not intended to be overwhelming, or to act as roadblocks to discussion – they are instead intended to open the gates to dialogue, to a creative brainstorming approach, to collaboration.

I know that my experience can't always be used as a working example – even for another student with a visual impairment, what worked for me won't necessarily work for the next person. My experience does, however, serve as an example, one that shows that integrating a person with a disability into a science lab is indeed possible – and if it can be done once, why not again?

Creating a Culture of Accessibility in the Sciences. DOI: http://dx.doi.org/10.1016/B978-0-12-804037-9.00017-6

When I started my Master's I didn't engage much in my own accommodation planning process – part of the reason for that was because the opportunity didn't present itself, but if I were really honest with myself, I also have to say that it's because I did not actively (or even passively, for that matter) involve myself in that process; or, frankly, seek out that opportunity I saw how … creaky, for want of a better word … my accommodations in the lab were during my Master's, and at the start of my PhD, resolved to fix that by engaging directly, with intent, in my accommodation planning. That was a much better, more effective decision in the long run – for one thing, I was able to make sure that everything stayed on track, that no agreements fell through, and that conversations happened when they needed to happen. I didn't realize it during my Master's, but people do get busy, and files drop to the bottom of piles. Things fall off the radar, so to speak – and none of us wants to be in that position.

Fortunately for me, I figured that out for my PhD. I was able to use the lessons that I learned in determining my postdoctoral accommodations, which themselves were an exercise in creativity.

The other thing I realized was that people don't often know the right questions to ask – there's a genuine willingness to help, but perhaps some reluctance to mess that up. It's really funny – the disability services staff are the accommodation specialists; the occupational therapists might know the adaptive technology; the faculty members know their courses, disciplines and labs. But I know my disability, and, as a result, none of us can have a real conversation without the others being at the table.

Introduction

Students with disabilities can, and have, worked "safely and effectively in the laboratory;" (Miner et al., 2001, p. 59). Unlike lecture or didactic instruction, where accommodations may be more consistent from course to course (e.g., note taking), accommodations for courses involving laboratory work tend to vary across disciplines (e.g., from chemistry to soil science and astrophysics to genetics) (Moon et al., 2012, p. 108), as well as disabilities. Furthermore, while accommodations may be similar to those provided to students in the classroom setting, significant differences are often likely, which entail intensive discussions and creativity to implement. This creates a unique challenge for accommodation planning, as a very clear understanding of the laboratory environment, task, tools, equipment, safety requirements, essential criteria, and disability-related needs must exist when determining accommodations and supports. Things are even more complex at the graduate research level, where the specific nature of the student's project, the nature of the research environment (type of lab or fieldwork setting), and the evolution of circumstances (project, disability, student-supervisor relationship) over time can have significant impact on the extent and type of accommodation required.

Unfortunately, as established earlier in this book, students with disabilities are significantly underrepresented in the sciences (see chapter 1: The landscape for students-with disabilities in the sciences) (Moon et al., 2012; National Science Foundation, 2013; Pence, Workman, & Riecke, 2003). Consequently, there is a lack of literature, best practices or a professional knowledge base on accommodating students with disabilities in the sciences, especially across disciplines. The material in this chapter has thus been designed to provide some context and offer a thought process for assisting decision-makers, in particular disability services providers, lab coordinators, and graduate faculty, in identifying accommodation strategies to ensure students with disabilities can participate in the sciences (Beckel, 2012).

In making science laboratories accessible to students with disabilities Sukhai et al. (2014, p. 9) noted "the importance of creativity in addressing academic accommodations, particularly with technology adaptations; the importance of a strong relationship, or "partnership," with faculty—either the course instructor/coordinator, or the thesis supervisor; the importance of a flexible teaching approach; and the importance of creativity in meeting the essential requirements for a course, program and discipline." Creativity and collaboration in identifying and implementing solutions to students' accommodation needs in STEM education and training are central, even axiomatic, concepts. These axioms resonate at all levels of STEM training in secondary education and the postsecondary system. They are particularly crucial at the graduate level with the central importance of the student-supervisor relationship (see the discussion in chapter 14: Faculty supervision of students with disabilities in the sciences).

Teaching practices, supports, and accommodations

This chapter outlines common academic tasks in science laboratories, both at the undergraduate and graduate level, and how various disabilities may impact the student and his or her performance of these tasks. We also discuss strategies and accommodations to guide decision-making regarding accommodations, supports, and teaching practices in science laboratories. As each student is unique, "different students and different disabilities will (therefore) need different accommodations. A single strategy will not be appropriate for all situations, neither will a single location in the laboratory be optimum for all situations" (Pence et al., 2003, p. 298). First, we focus on activities, disability-related impacts, and potential accommodation options in the context of the undergraduate or college teaching laboratory; later we offer a more general discussion of laboratory and fieldwork accommodations in the graduate setting. The subsequent chapters in this book focus specifically on technical assistance, simulation learning, technological accommodation, alternative formats, and physical accessibility and safety (see chapter 18: Human accommodation—laboratory/technical assistants in the sciences; chapter 19: Mainstream technology as accessibility solutions in the science lab; chapter 20: Assistive technology; chapter 21: Accessible formats in science and technology disciplines; chapter 22: Simulation learning; and chapter 23: Physical access in science laboratories).

Activities in a teaching lab setting in the sciences

Most courses with lab-based activities are modular; i.e., the length of the course is broken into modules, each lasting one or more session in the lab, where students are asked to carry out a series of integrated experiments toward a defined goal. Course materials for the lab may or may not include textbooks but usually include laboratory manuals developed by the teaching staff (Cook et al., 2009). Lab manuals are often provided as print copies and/or posted online on course websites or in online learning environments, for students to access. Furthermore, lab courses are often team-taught—the course coordinator often does not handle the day-to-day execution of the labs, but rather delegates this function to their teaching assistants (TAs). Students with disabilities must then make a determination about whether or not they should disclose their accommodation need to their TA, as they will interact most closely with that person. Complicating matters somewhat, many science lab courses have different TAs for different parts of the course. Students with disabilities must then navigate accommodation (and potentially disclosure, if they choose) to each TA in succession. Furthermore, due to the nature of group work in the teaching lab setting, students will also have to consider whether and how much to disclose to their peers, particularly those in the groups to which they are assigned.

A module often begins with a mini-lecture that outlines the theory behind the experiments to be conducted. The teaching staff may also demonstrate specific techniques for the students throughout the module. Students may be asked to work in groups to carry out the experiments involving the use of laboratory equipment, plasticware, and/or chemical and biological substances, based on the availability of materials and resources. Assessment of the module (i.e., measurement of the student's competency and knowledge in the lab environment) may take several forms, including a group presentation and a lab report. Assessment of the student's learning throughout the lab component of the course may take the form of quizzes, tests, bell-ringer exercises, and practical examinations.

Thus, in thinking about accommodation and inclusive practices in the lab environment, we have to consider the breadth of student engagement in the lab: by the instructor (e.g., through the taking of notes or the delivery of print materials, videos, and other forms of multimedia); laboratory demonstrations; the performance of group or team work; hands-on activities, including the use of technology (including lab equipment and assistive technology) and the use of chemical and biological substances; computer-aided data analysis and interpretation; and, assessment. Additional considerations in the lab environment include assistance and navigation, as well as physical access and safety (Curry, Cohen & Lightbody, 2006).

Accommodation in the graduate research laboratory

In the graduate research lab, students may be exposed to more safety hazards than at the undergraduate level—indeed, in an undergraduate teaching laboratory, especially dangerous items (e.g., radioactive substances, lasers, dangerous and potentially toxic compounds) are handled by the TA or instructor, or not used at all, for safety reasons. At the graduate level, the student may need to work with potentially dangerous devices or substances as part of his or her graduate research—or, working with his or her thesis supervisor, he or she may work around such exposure in designing the thesis project. Students may also be required to use protective shielding when working with certain substances (e.g., tissue culture in a biological sciences laboratory), and certain types of chemical reactions in a range of lab settings have to be carried using protected environments such as fume hoods, laminar flow hoods, or biological safety cabinets. For students with disabilities, this may be an issue due to dexterity, mobility, or visual challenges (Giesen, Cavenaugh, & McDonnall, 2012). As with all tasks in the lab setting, it is important to review what adaptation—equipment, assistive technology, lab assistant—may be the most appropriate to employ, and how best to implement it. If the student has appropriate lab accommodations at the undergraduate level—and those accommodations proved to be well-thought-out and successful—they can be used in the graduate setting.

There is a significant need for upfront preparation in defining and implementing the appropriate accommodation plan in the graduate setting, as was discussed in more general terms earlier in this chapter. There is also a strong argument in favor of the engagement of the student, thesis supervisor, and disability services staff in early and frequent conversations around accommodation needs, solutions, and disability management in the graduate setting in STEM. In an environment where the success of one's training is measured by productivity and research output, any significant length of time taken to establish accommodations is time lost. Careful planning early in the graduate research program ensures the time to implementation of an accommodation plan is minimized, and that questions of funding, ability to meet essential requirements, and required adaptations of the lab environment are able to be addressed with the participation of all stakeholders as efficiently as possible.

Accommodation in the fieldwork setting

In some disciplines, students may be required to travel to rural or remote sites for data gathering and collection. For example, students in forestry and environmental sciences may visit different biomes and ecological niches for observations of flora and fauna. Students in astronomy will visit observatories in remote locations. Students in geology, paleontology, archeology, and physical anthropology may visit dig sites around the world. For students with disabilities, travel to remote sites can be a logistical concern (Barker & Stier, 2013). Safety and ability to navigate successfully while in the field also assume greater importance, and to some degree, may be dependent upon whether the student is undertaking the fieldwork independently or as part of a larger group. Use of assistive technology in the field may be restricted by access to electricity, and assistive devices with long battery life or supplemental battery packs may be more appropriate in these contexts. Bulky or cumbersome devices are not as portable, and in choosing assistive technology solutions, the student, disability services staff, and faculty member should consider lightweight, easily transportable solutions.

Engaging a fieldwork assistant in a manner akin to a laboratory assistant (see above, and the discussion in chapter 18: Human accommodation—laboratory/technical assistants in the sciences) is often the most effective solution. Depending on the extent of assistance required, this individual may need some training in the discipline and be able to function effectively in the field under the direction of the student. As with a laboratory assistant, it is important that the student work well with the fieldwork assistant—indeed, if the student does work both in the lab and the field, whenever possible, it is best to have the same person fulfill these roles.

Finally, given the nature and time limitations of fieldwork, the student, his or her supervisor/faculty member, and the disability services staff along with a fieldwork assistant, if one is deemed appropriate, should carefully plan the activities scheduled for the length of the fieldwork. Contingency assessment and planning is also important, as is taking into account the length of time the student will need to perform tasks, with or without assistance. It is important to resist the temptation to attempt to accomplish too many tasks in the schedule alloted for the fieldwork, particularly if costs for travel and disability-related accommodations are significant—scheduling of fieldwork activities needs to effectively take into account the student's accommodation needs.

Discussing accommodations in the science lab or fieldwork settings

A good working relationship or partnership among faculty, the disability services provider (or accommodation specialists) and the student with a disability is extremely significant, and is a key element of supporting students with disabilities in science laboratories. Thus students should be encouraged to meet with faculty, instructors, and lab coordinators before the course or program commences and throughout the progression of the course or program to discuss procedures and should maintain contact throughout the term or program to ensure ongoing communication (Miner et al., 2001). It is also important to "visit the lab to make sure the area is accessible, learn about laboratory exercises beforehand, help faculty identify necessary accommodations, and help identify ways to fully participate" (Miner et al., 2001, p. 61). Ensuring that there is sufficient time for planning, design, preparation and implementation of accommodation solutions prior to the student's commencement of a course or program is vital. For the disability services provider, accommodations specialists, faculty/educators, and lab coordinators, advance notice of a student participating in a laboratory allows time to "research possible options for creating an effective laboratory experience…(and) it gives time for discussing options with the student (and faculty member), purchasing special supplies or equipment, and arranging for a laboratory assistant" (Pence et al., 2003, p. 295). Advance time for planning also allows the institution to be proactive in determining whose responsibility it is to pay for equipment (including Assistive Technology; see chapter 19: Mainstream technology as accessibility solutions in the science Lab and chapter 20: Assistive technology), a laboratory assistant (see chapter 18: Human accommodation—laboratory/technical assistants in the sciences), or other complex and potentially expensive accommodation solutions, before students commence their training period in a science laboratory.

This need for sufficient advance planning time also holds true for circumstances where it is clear that a student already participating in a laboratory environment may need significant modifications to his or her accommodation solutions, due to changes in program content, research direction and/or disability.

It is important to acknowledge the planning process required when addressing accommodations in the science learning environment. Accommodation specialists, faculty, educators, and students must consider what task(s) need to be accomplished, then consider what the World Health Organization classifies as the functional impairment that comes with the disability in question, in the context of the task(s) and environment the student is in. From those deliberations, all parties involved can decide, collectively or separately, what they believe the most appropriate response or accommodation to those impacts to be, and whether those accommodations or responses fit within the context of the essential requirements of the course, program, or discipline (see chapter 10: Essential requirements and academic accommodations in the sciences). For faculty and accommodation specialists, the model presented offers a framework for understanding and working through solutions for the needs of students. For students, the model provides a means of understanding how schools think about accommodation. Therefore the model provides a means by which all stakeholders in the accommodation discourse can speak the same language, and are able to evolve the most appropriate solutions (Gupta, Gelpi, & Sain, 2005). This thought process will be revisited as we move through the subsequent chapters in this section, looking at specific aspects of the accommodation dialog: human assistance (see chapter 18: Human accommodation—laboratory/technical assistants in the sciences); use of mainstream (see chapter 19: Mainstream technology as accessibility solutions in the science lab) or assistive (see chapter 20: Assistive technology) technologies; accessible formats (see chapter 21: Accessible formats in science and technology disciplines); simulation learning (see chapter 22: Simulation learning); and physical accessibility (see chapter 23: Physical access in science laboratories).

Conclusion

Reducing barriers to the participation of students with disabilities in science laboratories has unique challenges that can be met with creativity and an open mind. In this chapter, we demonstrated the wide range of options available for students in the sciences impacted by disabilities, some of which will be followed up on in subsequent chapters in this section.

Inclusive practices including reframing "learning goals, teaching techniques, materials, and assessment, individualizing for each learner with the help of flexible learning tools and media" can also support the participation of students in the sciences (Langley-Turnbaugh et al., 2004, p. 159). For the full realization of laboratory access for students with disabilities preparation and planning are key. Services providers should work with their science students to pay close attention to their program and degree requirements throughout their education in order to identify early intervention or engagement opportunities with faculty to effectively plan for the future (LoSciuto, Rajala, Townsend & Taylor, 1996).

Human accommodation— laboratory/technical assistants in the sciences

Chapter Outline

Introduction   209
The essential requirements argument   210
Does technical assistance provide an "unfair advantage?"   212
Does technical assistance provide the student "too much help?"   212
Is technical assistance "unrealistic?"   213
Does technical assistance mean the credit belongs to the assistant?   214
Is technical assistance too expensive?   215
Does the student gain the appropriate learning from having a technical assistant?   216
When to utilize technical assistance?   216
Finding an appropriate technical assistant   217
   Defining the role   217
   Job posting   218
   Interviews   218
   Training   218
   Evaluation   219
Conclusion   219

As much as I owe to my supervisors during my scientific training, I could not have been successful as a student and as a scientist without the assistance provided by three individuals during my Master's, PhD and postdoctoral training periods. These people, at various points of my training, served as my technical assistants in the lab setting, and helped me in very meaningful ways to successfully complete my training.

The initial idea for a technical assistant came from my Master's thesis supervisor – as with many of the accommodations I received over the years, I was somewhat uncomfortable with the idea to start with. Did I really need help to function in a scientific lab? Could I not do it on my own? As an undergrad, while in labs, I was always in a pair or in a group of three – and although I felt like a valued contributor to those groups, and was engaged in the work, I never had the chance to try something independently on my own. Such was the way of the undergraduate lab.

Creating a Culture of Accessibility in the Sciences. DOI: http://dx.doi.org/10.1016/B978-0-12-804037-9.00018-8

Graduate education was different.

Still, despite knowing and trusting my lab assistant during my Master's (in fact, I recommended her for the position), I felt like I had to figure some of this stuff out on my own. In molecular biology labs, to visualize the effects of various experiments on DNA, RNA, and protein, we do this thing we refer to as "running gels" – essentially a form of chromatography where nucleic acids or proteins are separated based on molecular weight and charge. In those days, we used to make gels manually (today, if you have the resources, you can buy premade gels) – that, believe it or not, was the easy part. It was, more or less, wet chemistry – just like cooking. Mix things in the right proportions, and do so in the right order and the right time, and you're good. The hard part for me was getting these small volumes of sample (1/100th of a milliliter) into tiny transparent holes cast into the gels called "wells." I have both contrast issues and hand–eye coordination issues, so imagine my success trying to put colourless liquid into a well cut into a colourless gel, resting in colorless solution. Even with creative tricks I picked up, this process of "loading the gels" would often take me longer than it took to actually run them and visualize the results.

It was frustrating, to say the least.

After a while, I began to wonder: "What's the point?" Not "What's the point" as in being frustrated and wanting to give up – but rather, "what's the point of me having to do this?" What was the value add to my training? I clearly knew (and still know, for that matter) how to load and run gels – but what mattered for me was the result the gel contained, not the loading and running the gel itself. So, what was wrong with me helping with all the prep work, but ceding the gel loading to someone else? After all, the original terms of my technical assistance accommodation – indeed, the terms that remained in place through subsequent iterations in different training environments and with different people – were that this person would function as my eyes and hands when needed.

When I started my doctoral training, my supervisor proved to be comfortable with technical assistance as an accommodation, but we refined the idea some. It had become clear during my Master's that certain things, such as radiation work, surgery, and work behind a physical barrier (or "shield"), were things I couldn't do, and so we factored those into the research project planning phase of my PhD– where we could, we ensured that safety issues would be avoided, and we also ensured that my new technical assistant (someone already in place in the lab before I got there) had the training needed for the rest of it.

It was then that, because of the scale of my project, and the complexity of the work, that I had to build a strong working

*relationship with my technical assistant, and that I honed my time
management and project planning skills. After all, I was responsible
for experiment development and planning, for troubleshooting,
and for data interpretation and analysis. I was responsible for my
assistant's time, and given the investment from my school, this wasn't
something to mess up.*

*So, we became a team, she and I, and worked well together for
the six plus years of my doctoral training.*

*When I became a postdoctoral scholar, challenges abounded in
maintaining this accommodation– for one, I was no longer a student,
and so the school could no longer pay for it. For two, few potential
postdoctoral mentors wanted to absorb the cost of a technical
assistant upfront. My eventual postdoctoral mentor hit upon a
brilliant idea – he'd recruited me well in advance of my start date
in his lab, and so we wrote grant applications together, building my
accommodation requirements (including technical assistance) into the
grant budgets. We were successful in obtaining grant funding and
paying for what I needed in the lab, which led to a very successful
three year partnership for that training period.*

*Since then, I have done more computational and supervisorial
work – my time as a wet bench scientist is over, not because I could
no longer do it, but because it was time to "graduate." Many scientists, at some point in their training and careers, end up leaving the
bench behind, and at the end of my second postdoctoral term, my
time had come. I carry with me still all the lessons I learned at the
bench, and all the skills I developed – both technical and transferable
to other settings. I would never have been able to succeed were it not
for the three technical assistants that I had, and their having served as
my "hands and eyes."*

Introduction

Human assistance in the scientific environment can take a number of forms, and
may, if applied effectively, be a highly beneficial accommodation for a student with
a disability, enabling—even empowering—his or her scientific skillset and training.
A human assistant can be someone who functions as the "hands and eyes" of a person with visual and motor coordination disabilities and is actively engaged in doing
an experiment with the student. A human assistant can function as a data collector,
taking measurements with instruction using instruments or technologies that are not
accessible to a student with a disability, in the lab or in the field. A human assistant
can function as a scribe for someone who needs help taking notes or keeping records
in the experimental setting by writing down comments, notes, or analyses dictated by
the student. The functions an assistant performs depend on the specific needs of the

student in the context of his or her discipline, disability, and environment. However, human assistance in the scientific setting is highly debated, which acts as a roadblock to the application of this approach.

Can science be done with help?

For students, faculty, and services providers alike, this is the question that most often gets raised when considering the potential impact and implications of human assistance in the scientific laboratory and fieldwork settings. Having a "lab partner" or a "technical assistant" or a "fieldwork buddy" provide assistance or support as an accommodation in an experimental setting—particularly during a student's training while in postsecondary education—is sometimes seen as an unfair advantage over his or her peers. This type of accommodation can also be viewed as unrealistic or impractical in the "real world" and be dismissed as inappropriate, or as an "undue hardship." Students may turn down the opportunity for "help" thinking they need to complete tasks on their own—and faculty, during training, can reinforce this idea by putting students in isolating situations, where they are not encouraged to seek help from others in their research group. Human assistance can also be seen as having to "share the credit" between the student and the assistant for the work done. Alternatively, the assistant may be seen as deserving of the credit for the work, as opposed to the student, since the student may have not physically performed the tasks in question. Human assistance can also be expensive, depending on the training and skillset of the assistant, and how many hours a week the assistant works with the student (Miner et al., 2001). Finally, some argue that a student will not gain the appropriate training if a human assistant is working with him or her.

In this chapter, we will apply the essential requirements argument (see chapter 10: Essential requirements and academic accommodations in the sciences) to human assistance as an accommodation in the science lab and during fieldwork, and deconstruct the arguments against human assistance. We will examine scenarios where human assistance is beneficial, and consider the processes that need to be implemented to ensure the appropriate utilization of this type of accommodation. In this chapter, we will apply the essential requirements argument (see chapter 10: Essential requirements and academic accommodations in the sciences) to human assistance as an accommodation in the science lab and during fieldwork and deconstruct the arguments against human assistance.

The essential requirements argument

Let us return, for a moment, to the essential requirements concept (see chapter 10: Essential requirements and academic accommodations in the sciences), and the issue of defining the required competencies for a discipline, program, or field, versus the measurement of those competencies. The reasoning around human technical assistance as an accommodation goes something like this: If a student is required to demonstrate a specific competency in his or her field, and this task requires help in order to complete it, then the student has not demonstrated that competency effectively. If the student cannot demonstrate that competency without accommodation, and it is something that is crucial to his or her being able to effectively perform as a scientist in the chosen field, then he or she cannot, by definition, be successful in that field. Therefore the accommodation is not permissible—and, if there is a safety issue or patient care issue attached to this conversation, the student may not be permitted to continue in the course, program, or discipline.

The logic behind this argument is sound—indeed, this is exactly how the essential requirements argument should be applied. However, the real challenge is that while many professional disciplines aligned with the sciences (e.g., medicine, nursing, pharmacy, dentistry, and other related health professions, as well as the engineering disciplines) have accreditation bodies that have invested time and effort in identifying essential requirements for their fields, many basic and applied sciences have no such definitions in place for training programs. Alternatively, even if such requirements exist for a given school, they are not applied universally (Oakley et al., 2012). Put another way, a student training at Cambridge or Oxford may encounter different perceptions of the essential requirements of his or her field, compared to a student at Harvard, or at a school in India or China. Furthermore, a student can encounter different perceptions of the essential requirements of his or her field from faculty members across the hall from one another in the same department, complicating the issue exponentially. The implication of these differences is not the manner of application of the essential requirements argument to human technical assistance but rather the *appropriateness* of its application.

An example may help illustrate this argument. In the biological sciences, a student may be using an animal model of disease to study that disease. If that student has a disability affecting hand–eye coordination, he or she may require human assistance to perform surgery on the animals. Some faculty could argue that not being able to perform surgery independently means that the student does not meet the essential requirements of the field. This would be true if the field were relevant to physiology—if the student were studying this animal model in the context of a veterinary program, or anatomy, or physiology, for example. However, if the student were in a molecular biology-based program, where the use of the model was incidental to the data derived from the model, the fact that the student needs help carrying out animal surgery is irrelevant. In that program, the student needs to demonstrate that he or she knows how to effectively interpret and analyze the data resulting from the surgery.

Now let's consider another example, this time from the physical sciences: Does a student need to physically construct an electrical circuit to understand the fundamentals or practical applications of current, voltage, and resistance in circuits connected in series and in parallel? Arguably, knowing how to build a circuit by hand can help in the learning and mastery of the competency, but that has more to do with how a student may learn, which is a (mostly) separate conversation from one about that student's disability. Furthermore, knowing how to build a circuit itself demonstrates a different competency than understanding current, voltage, and resistance. Again, understanding current, voltage, and resistance helps in being able to build a circuit and so, although the two competencies are intricately related, even intertwined, they are not identical, and should not to be conflated in our thinking (Roberts, 2013).

In that context, if a student needs to know how to build a circuit by hand, technical assistance to build that circuit is not an appropriate accommodation. If building that circuit is but a means to an end—i.e., it is a way to measure a different competency, which could be evaluated differently—then technical assistance is a viable accommodation strategy (as would be simulation learning, to which we return in chapter 22: Simulation learning). Thus rigorous and appropriate application of the essential requirements argument, rather than outright forbidding human technical assistance, permits the use of technical assistance (or simulation learning, as appropriate) in the right circumstances, which need to be carefully defined in the context of the program, discipline or field (Roberts, 2013).

Does technical assistance provide an "unfair advantage?"

Giving students with disabilities an "unfair advantage" is a systemic concern expressed by faculty in the sciences around accommodation generally and toward technical assistance specifically. The idea is articulated like this: "You, the student with a disability, through the accommodation process, have access to a resource your peers do not have. Therefore you have an unfair advantage over your colleagues." This notion is predicated on an often-incorrect assumption—that the student with a disability starts at the same place and in the same circumstances as his or her peers.

Accommodation, as we learned earlier in this book, is intended to "level the playing field" for students with disabilities (see chapters 1, 3, 10 and 17: The landscape for students with disabilities in the sciences; Barriers faced by students with disabilities in science laboratory and practical space settings; Essential requirements and academic accommodations in the sciences; and Students with disabilities in science laboratories—teaching practices, supports, and accommodations)—put another way, accommodation is implemented to correct for disparities that work against students with disabilities (Rose et al., 2006). If technical assistance has been deemed to be a viable accommodation strategy, this decision has been taken in order to provide the student with a disability the opportunity to perform in a science-based environment at the same level as his or her peers. Rather than providing an unfair advantage to the student, technical assistance when implemented appropriately and in the context of the essential requirements of the field, corrects to varying extents an existing disadvantage.

Does technical assistance provide the student "too much help?"

The issue of providing the student "too much help" is associated with the challenge of "unfair advantage" just discussed. There is no way around the notion that technical assistance, by definition, provides the student with a disability help in the science-based environment. How much help then is "too much?" If there is such a thing as "too much help," how can this be corrected for in accommodation planning, implementation, and evaluation?

To address this question, we start with what the duties of the technical assistant are, as defined in the context of the student's disability and accommodation need, as well as the essential requirements of the program and discipline. Planning, executing, and interpreting the results of experiments—i.e., playing in entirety the role of the student, for the student—would be clearly too much help. Likewise, carrying out writing of manuscripts, computer code, or technical documents that are a part of the student's work in the research group would constitute too much help. However, handling materials deemed too dangerous for the student with a disability, carrying out tasks under the direction of the student that he or she cannot execute physically, taking measurements the student cannot physically take under the direction of the student, and scribing for the student based on dictation, among many other duties, are all appropriate duties of a technical assistant.

There are two key points that must be highlighted. The first is that the technical assistant works under the direction of the student; the second is that the technical assistant does not assume any of the critical thought aspects of the student's role in the research group (i.e., planning and design of an experiment, code, or manuscript; data interpretation and analysis; data synthesis; or, troubleshooting). If the technical assistant assumes critical (or analytical) thought roles that should be undertaken by the student, several consequences ultimately become visible. Most glaring of these is when the student demonstrates inadequate knowledge of and preparation for what is supposed to be his or her own work and is measurable in conversation. Faculty and services providers seeking some precedent with respect to the definition of "too much help" on the part of technical assistants can look toward guidelines offered to tutors engaged as part of the accommodation plans of students with learning disabilities, which address this issue in a different context.

Is technical assistance "unrealistic?"

One argument against the provision of technical assistance as an accommodation is that such an approach is not feasible in any setting outside of the undergraduate or graduate training environment. Thus if the student wishes to be an experimental scientist but is unable to perform the bench science unaided, he or she would be unsuccessful in his or her career. This rationale is correct in situations where the student has a desire to become an active bench scientist (a lab technician, e.g., or a technologist in the allied health disciplines). In such a setting, the employee is required to work unaided at the lab bench, and technical assistance defeats the stated purpose of those employment positions.

It is worth noting that many science-related occupations do not involve any work at the lab bench or directing others in those settings. In these types of scenarios, the student will need to demonstrate understanding of the work being conducted at the bench, without having to actually perform that work themselves. Students in science programs that eventually lead away from the bench are in positions where technical assistance would not be harmful to their learning and demonstration of essential competencies (Sharby & Roush, 2009).

Furthermore, in situations where the technical assistance required by the student or employee is passive in nature, as opposed to active (i.e., gathering measurements, as opposed to physically assisting with experiments), employers may consider this a realistic approach. Indeed, that level of assistance may be attainable without formal accommodation, via help from other members of the team the person with a disability is on.

Thus rather than all forms of technical assistance in all settings being unrealistic for employers to provide, the reality of the situation is much more nuanced. While there are some circumstances where technical assistance is unfeasible—indeed, even inappropriate as an accommodation—there are many others where it at least warrants consideration in the dialog with the student or employee as a potential accommodation strategy (Sharby & Roush, 2009).

Does technical assistance mean the credit belongs to the assistant?

The intellectual property argument is another objection raised against the involvement of technical assistants in accommodating students with disabilities in the sciences. Students may object to what they view as an unfair sharing of credit. Perversely, faculty may object on the grounds that the intellectual property (i.e., the data collected, or the manuscript transcribed) should actually belong to the assistant, not the student.

The defense against this argument again rests with the duties of the assistant, as defined by the accommodation need and the essential requirements of the discipline or program. As long as the role of the assistant does not include aspects of the critical thought process that is crucial to the student's training, there can be no intellectual property or ownership question. The sharing of credit in this circumstance is entirely up to the student and faculty member(s) involved, and will depend very much on the culture of the field, department, and faculty.

Another side to this argument can be demonstrated in a comparable scenario. Imagine that a scientist makes a groundbreaking discovery by using technology from a scientific supply company. One might argue that the scientist in question did not develop the technology that was used, but rather the protocol for the specific experiment. Should the scientific supply company have a claim to royalties or the patent? Is the scientist obligated to give or share a portion of the credit in subsequent publications? A technical assistant is a vital and valued member of any research team, but this does not place them in a position to receive credit anymore than a traditional research technician who has been hired for data collection or bench work. If these stipulations are made clear as part of the recruitment, interview, and hiring process, the issue of credit sharing may be avoided altogether.

Is technical assistance too expensive?

Often, making arrangements for technical assistance is a new experience for the student, faculty, and disability services staff. There are many ways to implement this accommodation approach, depending on the needs of the student and the environment he or she is working in. According to disability services providers, the majority of lab assistants are volunteers with some background in the sciences or are peers already in class. For example, students often work in pairs in the laboratory, so it may be appropriate for the lab coordinator to give the student with a disability a partner to assist with specified tasks. However, disclosure to the partner may not always be necessary. In some cases a lab supervisor may assign tasks to pairs, keeping in mind the needs of a student with a disability.

If a peer helper serves as the assistant, his or her needs must also be considered. The helper's learning should not be compromised. One of the drawbacks of this arrangement is that the peer may not always attend class. Considering this possibility in advance with all students in pairs may help reduce reliability issues. At times, it may be appropriate to hire a lab assistant, especially if the level of assistance required or the technical skillset needed is significant (e.g., 5 hours a day, full days for graduate students; specific lab skills such as surgery, handling radiation, or other specialized competencies, etc.) or where the learning of the peer may be compromised. In some cases, disability services providers have also reported that teaching lab supervisors had served as such supports for students. This type of situation is often dependent on variables such as personality, enthusiasm, and availability. These unreliable variables often make the applicability of this option unpredictable.

The most expensive form of technical assistance accommodation is a full-time dedicated technician in a research lab (e.g., graduate or postdoctoral) setting. The cost of this type of accommodation depends on institutional rules, etc. Senior technicians with more specialized skill sets would be more expensive than more junior technicians. Funding these positions requires creativity and commitment on the part of the student's supervisor, the department, the disability services office, and the institution (Sukhai, Mohler, Doyle, & Carson, 2014). There are ways to ensure successful implementation of this accommodation to benefit the student over the short and long term— e.g., specialized funds from within the university or college or planned inclusion of the salary of the technical assistant on grants written by the supervisor—but they all require collaboration, dialog, and creativity to implement and monitor.

As with most accommodations, the answer to the question "is technical assistance too expensive?" is that it depends very strongly on the student's needs and the environment. Is the student in a teaching lab setting? Fieldwork? A research lab? What tasks are required to be performed by the technical assistant? Will assistance from the faculty member/lab supervisor suffice? Or from a peer helper? Is dedicated assistance required, and if so, for what fraction of full-time equivalence? No two situations will be the same, and each student's needs should be evaluated with careful attention to his or her specific circumstances and program/discipline requirements. Of note, the funding source (e.g., department, central funds of the postsecondary school, partnership with disability services office, etc.) and rate of pay (e.g., casual rates, TA wages, etc.) for the assistant should be determined well in advance.

Does the student gain the appropriate learning from having a technical assistant?

This argument against technical assistance as a viable accommodation strategy is related somewhat to the issue of what kind of help the student is getting. For the student to receive assistance without interfering with the critical or analytical thought processes he or she must demonstrate in order to fulfill the essential requirements of his or her course, program, or discipline, the student must be intimately involved in the planning, execution, interpretation, and (if necessary) troubleshooting of experiments. The student is ultimately responsible for efficient use of the assistant's time, and to be effective as a tandem with the assistant. Anecdotal evidence from students with disabilities who have utilized assistants in a variety of setting indicates that, rather than knowing less than their peers, or being less engaged in the learning process, these students have to be *more* prepared, *more* knowledgeable, and *more* engaged with all aspects of their learning than their peers. Because the assistant works at the direction of the student, the student must have thought through the process or task to be carried out, as well as various contingencies in case things go wrong. This level of advance planning by the student requires a deep theoretical and practical knowledge of the tasks being performed, as well as a level of contingency planning expertise, that many of the student's peers would not be exposed to. Arguably, therefore, a student with a disability who is working with a technical assistant gains a deeper level of learning than his or her nondisabled peers.

When to utilize technical assistance?

Having addressed the arguments against technical assistance as an accommodation for students with disabilities, we turn our attention to the appropriate circumstances in which technical assistance may be warranted (Sukhai, Mohler, Doyle, & Carson, 2014). Again, depending on the student's accommodation need and the essential requirements of the discipline or program, technical assistance may be warranted in order to:

- Assist the student in lifting, moving, and/or carrying materials he or she is physically unable to lift, move, or carry;
- Assist the student in performing manual tasks that require significant hand–eye coordination;
- Assist the student in performing tasks that may compromise his or her's safety in a lab setting (e.g., handling radioactive materials);
- Assist the student in using instrumentation he or she is required to use, but unable to utilize without assistance;
- Assist the student in recording data;

- Assist the student in record keeping and transcription of lab notes, manuscripts, and/or other technical documents; and,
- Assist the student in other lab-related tasks that are required to be performed, but that he or she needs help with.

Technical assistance should be designed to aid the student's training and development, without compromising learning. The relationship between the student and the technical assistant also needs to be strong; in addition to being able to get along as individuals and as a team, the student needs to be able to trust the technical assistant without worrying that the work will be compromised. Although the student may prefer to work with one lab assistant or one fieldwork assistant throughout his or her entire academic career, depending on the extent and nature of his or her training, the student will likely need to work with a variety of assistants employed on short-term contracts. The following sections describe some best practices for finding and retaining quality technical assistants.

Finding an appropriate technical assistant

Defining the role

The requirements/responsibilities of the assistant should be defined in consultation among the student, course instructor or faculty program coordinator, and the disability services staff. The student with a disability must also understand the expectations and his or her role. The student will direct his or her own activities with the support of the lab assistant. Specifically, it is important to:

- Clarify the learning objectives of the course/program and what assistance the student will require to meet the learning objectives.
- Clearly outline the role of the assistant.
- Clarify boundaries (discuss the integrity of output; what the student must be able to do independently).
- It is also important for faculty members and disability services staff to keep in mind that the technical assistant is not being hired as a disability specialist or special education facilitator. The technical assistant is hired for the research tasks at hand and should be chosen based on well-suited personality traits and scientific background, skills, and experience.
- There is a lot to be gained for a technical assistant. In addition to gaining references and research experience, the technician is becoming familiar with many technologies, developing the ability to think critically in new situations, and adding the component of ingenuity to his or her repertoire of lab skills. A technician who is still developing his or her career may prove to have the flexibility needed where perseverance in the face of uncertainty is vital (Sukhai, Mohler, Doyle, & Carson, 2014).

If the faculty member working with the student is not directly involved in the recruitment process, it is important to ensure that the course instructor/faculty member is aware of the assistant's role and that the assistant does not affect communication between the student and instructor. Finally, it is important to consider any existing preferences on the part of the student or the faculty member. For example, consider whether there is a preference for a female or male assistant (e.g., this may be a significant consideration in some fieldwork-oriented circumstances), or whether there is a preference for someone already part of the research group.

Job posting

A job posting should include a description of the duties, the type of assistance required, the working environment and, with the permission of the student, reference the type of disability the student has. The job posting should also include any required skills, technical expertise, knowledge, or personality traits that are desirable for the position (e.g., demonstrates great patience, presents innovative ideas to ensure inclusiveness of activities, etc.). The student should have a role in the preparation of the job posting. The job posting can be circulated throughout the faculty member's and institution's networks, to other schools with similar programs, to organizations in the field of study, and throughout disability-related networks.

Interviews

Ideally, the student should participate in the interviews and help choose the final candidates. Working together, the student, faculty member, and disability services staff (if appropriate) should develop a list of interview questions that will specifically help to determine the appropriateness of each candidate for the position. It is best for the student to briefly speak about his or her disability at the beginning of the interview and to explain how it relates to the role of the assistant. For example, "I am blind but when I was younger I had a little vision and I could see colors; when you are describing things to me, telling me the colors is very helpful." The interview process should also include a test to help determine which candidates best fit the position. Let's take a look at an example of a university looking for an assistant to work with a blind student. The interview process includes requests that candidates verbally describe an image/graph to the blind student. Candidates are also given the opportunity to take part in a sighted guide demonstration. If the student will be working with the assistant in-person, the interviews should be conducted in-person so that the student can best determine which candidate will best fit the position. Following the interviews, the student, faculty member, and disability services staff (if appropriate) should narrow down the list of candidates, conduct reference checks, and then proceed with making a decision. It is a best practice to keep the resumes of the top candidates in case you need to contact them as back-up assistants or to replace an assistant who has moved on.

Training

Training should be provided to help the assistant better understand his or her role and also provide the assistant with any disability awareness training needed. For example, for a blind student, the assistant should be provided with sighted guide training and information about working with someone who is blind. Training regarding safety in the laboratory may also need to be arranged if the assistant does not have previous laboratory experience. If the accommodation plan includes specialized technical assistance, the assistant will likely need to complete the institution's required training and certification courses.

Evaluation

The faculty should check in at the beginning, middle, and end of the assistant's work term (these checks should be done annually or semiannually if the assistant is working long term for a graduate student), so the student can provide feedback and also receive feedback from the assistant. This process will help to establish a working relationship that fosters open and honest communication, where both the student and the assistant feel comfortable discussing what is working, what is not working, and what should be done differently.

Conclusion

One of the accommodations for students with disabilities in science programs is a lab assistant, field work assistant, or other type of scientific academic assistant. An assistant may be required to pour or carry chemicals from one place to another, handle equipment, help navigate through the laboratory, describe material, etc. While there are some significant perceptions and arguments against the employment of technical assistance in accommodating students with disabilities in the sciences, a careful evaluation of the student's accommodation need in the context of the essential requirements of the program, discipline, or field is the most objective method to determine whether technical assistance would be of benefit. For technical assistance to work well, the student, faculty member, and disability services staff need to be fully engaged in the recruitment, retention, supervision, and evaluation of the assistant. Indeed, when applied appropriately, technical assistance can be highly beneficial to the student and in some ways forces a student with a disability to work differently than peers to achieve his or her goals.

Mainstream technology as accessibility solutions in the science lab

19

Chapter Outline

Introduction 223
Case Study #1: Using robotics to handle small volumes in a
   biological sciences research lab 224
Case Study #2: Microscope slide scanners in anatomy,
   histology, or physiology teaching labs 224
Case Study #3: Universal design in the lab—computer-aided
   instrumentation in a physics laboratory 225
Case Study #4: Adaptive technology brought mainstream 225
Case Study #5: Mainstream technology adapted for accessibility 226
Case Study #6: Accessible mainstream technology solutions—build
   with universal design principles in mind 226
Case Study #7: Low-tech solutions 226
Determining the best technology solution for a student with a disability 227
Student considerations 228
Engineering custom technology solutions: the need for a knowledge base 228
Conclusion 229

As part of my doctoral thesis project, I needed to examine cells taken from the bone marrow of mice under the microscope as the first test to determine if these mice had leukemia.

There was only one problem.

I can't resolve anything looking down a microscope eyepiece.

We had to come up with a solution in reasonably short order. Fortunately, I knew going in what my functional limitation with a microscope was, and once I had figured out what needed to be done with these mice, I realized we'd have a challenge. The first thing I did was let my supervisor know, and I suggested that we would have to engineer a solution, or at least see if there was something commercially available that could do what I needed. I was open to all possibilities at this stage, and was willing to work to facilitate my own solutions.

After I got her permission to go ahead and do my research, and I got carte blanche from the disability services office to figure out what would work best – after all, as accommodation specialists, they

Creating a Culture of Accessibility in the Sciences. DOI: http://dx.doi.org/10.1016/B978-0-12-804037-9.00019-X

knew assistive technology well, but had little background in biology, microscopy, or engineering, and couldn't do more than advise me in evolving a solution – I got down to work.

I researched microscopes on the market, and trialed various microscopes within the research institute – the latter to see if there was a solution already on hand that I could borrow when needed. I spoke to sales reps. It turned out that that there wasn't anything on hand that would accomplish exactly what I required.

This was 2001, and computer-aided microscopy – now a staple of the field – was in its infancy then. I'd figured that what I needed was a microscope with a video camera that could connect to a TV monitor – that way, I could look at the TV, without having to look down the eyepiece, and still be able to see everything in the field of view of the microscope lens.

I didn't think that solution was too hard to implement – after all, all the technology was off-the-shelf. The problem was that, apparently, no one else had thought of solutions like that. The closest we came was a microscope with a computer-controlled camera – but it still required you to look down the eyepiece to identify the image you wanted to take a picture of.

(This doesn't diminish by any stretch anyone's intelligence – I note it to point out one of the very interesting things about working in science, technology, and engineering: A whole range of solutions are possible to a given problem, but until someone brings a specific perspective to the table, some solutions are not considered because they are outside others' range of experiences. Perhaps this is a new way to think about diversity in science....)

Anyway, the standard solution that was available at the time was a poor substitute for what I needed – for one thing, it meant I couldn't use the microscope independently, which was what I was really angling for. Second, it was a brutally expensive solution, and I wasn't sure I could justify it to the disability services office, because it was a poor solution at best that didn't meet all our objectives.

Then one day, I spoke to a sales rep from a big company who stopped by. I explained my challenge, and told him what I was looking for. His response: "Seems like something we should be able to assemble for you – let me talk to our engineers." A few weeks later, he got back to me to say that they had worked out a strategy and were field-testing it, but that if it worked, we'd be able to buy a system that could do everything I required.

Later in the year, we took delivery of a microscope with a camera with high resolution that connected to a TV monitor. A cathode ray tube TV monitor, but this was 2001, after all. The important thing was that I could get images projected from the microscope stage to the TV, in real time, and at high quality and resolution.

Even better: It was three times less expensive than the next best solution.

(An alternative solution to the one I utilized is presented in the second case study in this chapter; of note, this technology did not exist when I was in my doctoral program.)

That microscope was in service for almost a decade, through my PhD, and the graduate degrees of several other students in the lab, as well as for the length of my first postdoctoral fellowship. We decommissioned it when the research lab was decommissioned, in the summer of 2012.

That one piece of custom technology defined in many meaningful ways the first half of my PhD– I used it practically every day, and it was a crucial tool in gathering the data that formed the foundation of my thesis project, but it wasn't something we could just buy. Off-the-shelf parts, assembled with creativity and innovation, put to work as a technological solution to a real and present accommodation need.

Introduction

Many learning needs of students with disabilities can be accommodated with the creative adaptation of mainstream and off-the-shelf technology or equipment. Indeed, adaptation of existing or readily available scientific equipment, or of newer technologies such as smartphones and tablets, can open up a range of accommodation possibilities in teaching and research lab settings and during fieldwork. In considering what technology solutions may be appropriate, it is important for the student, faculty member(s), and disability services staff to sit together—preferably with a technology expert—to discuss what task(s) the technology solution is intended to aid, the range of possible technology solutions, and what may likely be the best fit for the student based on his or her specific accommodation need(s). It is worth noting that any adaptation of off-the-shelf or mainstream technology for the purposes of accommodating a student with a disability is merely the utilization of a tool for a different purpose than it was originally conceived. The essential requirements argument discussed throughout this book applies again—the adapted technology solution is merely a tool for the student to use in achieving the appropriate competency, and should not be misconstrued as a method to shortcut or interfere with the student's learning (Roberts, 2013). Additionally, it is important for faculty and disability services members to be prepared for varying degrees of trial and error during these processes. As an idea evolves, the needs of the student are likely to evolve as well. This is all part of the process of adaptation and provides vital learning opportunities to all who are involved. Such occurrences should be embraced and taken advantage of.

The range of possible technological solutions using off-the-shelf or mainstream technology is enormous; rather than attempting a comprehensive listing of possibilities, we provide some illustrative case studies, alongside some suggestions on how best to evolve the most effective mainstream solutions and some student challenges to consider when deciding how and when to use such solutions.

Case Study #1: Using robotics to handle small volumes in a biological sciences research lab

Many molecular biology applications require the handling of very small (often less than 1/100th of a milliliter) volume of solutions in setting up experiments. For persons who have difficulty with low-contrast or small objects, or who have fine motor disabilities, handling small volumes such as this can be challenging. Many universities and research institutes have dedicated core robotics facilities that allow for the effective handling of small volumes in a timely manner. One challenge with using these facilities is that they often work best in a "high-throughput" setting—i.e., the robotics are set up to handle up to 384 samples at a time. This issue can be resolved with careful planning on the part of the student, in terms of experimental setup and management; rather than restricting or minimizing the complexity of the study design for manual implementation, a more wholesome study could be designed instead. Alternatively, two or more parallel experiments could be run simultaneously, with effective preparation. Depending on the nature of the student's project, judicious use of a core robotics facility may be a more effective accommodation than a part-time or full-time technical assistant, and offers significant savings in time and scalability of the student's experimental plan. Indeed, access to a core robotics facility is often of benefit to all students and research trainees within the university or research institute. Access to this resource for a student with a disability maintains a level playing field for the student in comparison to peers.

Case Study #2: Microscope slide scanners in anatomy, histology, or physiology teaching labs

Some students with visual disabilities have difficulty focusing or resolving images in a microscope field of view. For such a student, being able to visualize and interpret the information contained on the microscope slide is of paramount importance to course or program essential requirements, as opposed to the operation of the microscope itself. Students with visual disabilities in a teaching lab setting can, in this scenario, take advantage of specialized scanners capable of imaging a microscope slide. While this technology is too expensive for teaching labs or individual research labs, it may be available as part of a core microscopy imaging service at the university, or at a nearby teaching hospital or research institute. Slide scanners render high-resolution images as graphics files that can be viewed by a student on his or her laptop or tablet, where they may choose to, as necessary, magnify the image, zoom in and out, change contrast, and otherwise enhance the image to enable them to view it effectively and gather the appropriate data and observations. Additionally, rather than using camera equipment attached to the microscope in the lab to take photographs of the slide, the images from the slide scanner can be used for inclusion in the student's lab report or data records.

Case Study #3: Universal design in the lab— computer-aided instrumentation in a physics laboratory

Curry et al. (2006, p. 34) argue that "some of the best inquiry-based learning opportunities in science authentically simulate lab and field investigations undertaken by active researchers and scientists." Universal design guides the identification and selection of tools and equipment—including those used to prepare experiments and gather and visualize data—to ensure participation of almost all learners while minimizing the need for individual accommodations. These tools can incorporate multimodal representations of the information they are designed to deliver, increasing the opportunity for diverse learners to accurately capture the evidence and data collected. These tools can also enable the collection of data in alternative formats or using different methodologies, ensuring the accessibility of experimental results to all students.

Computer-controlled laboratory equipment is one of many examples of universal design tools (Burgstahler, 2012). Today, a number of laboratory devices, which are required for the majority of experiments in most fields, are computer-controlled and therefore likely accessible to a diverse group of users. For example, in a chemistry lab, a sensor (or other interface hardware such as a temperature probe) is connected to a computer or handheld device, and specialized software controls the timing of measurements and the recording of data throughout the experiment. Because the data are recorded and displayed digitally, access can be customized based on learner needs and preferences.

Case Study #4: Adaptive technology brought mainstream

It is also possible to retrofit traditional assistive technology solutions (e.g., a closed circuit television, or CCTV) to serve specific purposes in the lab environment for students with disabilities. One example of this could be the use of a CCTV as a dissection platform for animal surgery in a biological sciences lab, or as a platform on which to mount items requiring fine motor manipulation in a physical sciences lab, where magnification would be beneficial for the student. It is thus important for faculty to be aware of available assistive technology resources, and to work with the student and staff from the disability services office on campus to determine the most applicable and creative uses for these technologies.

Case Study #5: Mainstream technology adapted for accessibility

iPads and iPhones as well as Google and Android tablets and smartphones are versatile devices with a range of accessibility features for people with disabilities. These devices also have advanced optics and higher screen resolutions—in many cases, the camera in a recent-generation iPad or iPhone has higher resolution and is of better quality than the camera in a CCTV. Because of the range of Bluetooth accessories for these devices, the virtually infinite number of scientifically relevant apps, and their long battery life, they are ideal as data-gathering tools in the field or in the lab, and can easily be adapted for accessibility. For example, the camera on a smartphone or tablet could be used as a high-resolution magnifier; the voice dictation features, with the right training, can be used in place of a scribe for a student with learning disabilities or carpal tunnel syndrome, or as a data recorder for all students. For students with visual disabilities, adjusting the contrast settings and font sizes on a tablet can yield significant benefit in reading primary source articles from scientific literature. Enabling speech-to-text features on a tablet or smartphone can ensure that a hard-of-hearing student can follow a short presentation. Students, as direct users of technology, are in the best position to determine what potential applications or solutions work best for them.

Case Study #6: Accessible mainstream technology solutions—build with universal design principles in mind

Several companies have recently started developing instrument solutions for the science lab environment that incorporate universal design principles. For example, digital balances and pH meters with large screens and big, easy-to-use controls are more accessible than devices with small screens, fonts, and controls, and a digital pipette is more accessible than an analog one. Other companies, such as Independence Science, based in the United States, have developed accessible versions of instrumentation and software previously not accessible to a person with a disability. As more science equipment is designed with accessibility and universal design in mind, and as more labs adopt these devices over older, more inaccessible versions, the costs for these instruments will begin to drop. While these instruments are accessible to people with disabilities, they are useful to all members of the lab, and they can prove to be cost-effective solutions to incrementally enhancing the accessibility of the science lab environment (Burgstahler, 2008).

Case Study #7: Low-tech solutions

Depending on a student's accommodation needs, a variety of low-tech solutions may be used in the laboratory setting to better ensure his or her successful integration in the lab. For example, a student with a sight impairment may benefit from having access to magnifiers built for the science lab, or from desk lamps mounted over his or her bench for better lighting. A student with chronic back pain may benefit from an ergonomic chair, as opposed to a laboratory stool. These low-tech solutions can be affordable and easy to implement, yielding significant benefit for the student with a minimum of effort. Students often have a sense as to what has worked for them previously in other lab environments, and may be able to effectively advise their accommodation specialist or faculty member as to potential options.

Determining the best technology solution for a student with a disability

Identifying the best mainstream technology solution for a student with a disability needs to begin with the student's specific accommodation need in the context of the discipline, program, or course and the nature of the task(s) the student needs help to perform. The nature of the environment the student is working in also needs to be considered—is the student in a teaching lab, a research lab, or the fieldwork setting? Once the group—student, faculty member(s), disability services staff, adaptive technology expert—have agreed on the accommodation need, tasks, and environment, consideration should be given to whether existing equipment in the lab, research institute, department, or university can be adapted to aid the student, or whether a smartphone or tablet can provide a cost-effective and accessible solution (Sukhai, Mohler, Doyle, & Carson, 2014). If some combination of those two possibilities can provide the appropriate solution for the student, the group should proceed to field test the solution and ensure, with the student's input, it is appropriate and effective.

If there needs to be a bigger investment in new equipment for the lab, or modifying adaptive technology or software, the group needs to work through potential technology solutions using off-the-shelf resources and engaging various other experts, as needed, from within the university or research institute, or the private sector (e.g., scientific equipment supply companies). It is likely that the idea(s) the group develops would require some level of innovation to customize and implement—as such, with the student's permission, it is important to record this knowledge for future application. After potential solutions are identified on paper, some consideration should be given to how best to fieldtest the options to choose the best one. Funds for the purchase of the equipment also have to be found, and usually come from the student's home university, through the supervisor's grants, or through financial aid opportunities available to the student.

It is important to recognize that the student's accommodation circumstances are likely to change—the student will finish the class, and may need to take the accommodation with them to the next learning setting, if feasible to do so; the student's project in graduate school may evolve or change completely; or, the student's disability may likewise evolve over time. Given these considerations, it is important for all members of the group to meet regularly to review the successful implementation of the accommodation, and whether new or additional pieces of equipment may be needed by the student.

Student considerations

Some students with disabilities are reluctant to broadcast to their peers what they might consider to be "special status" associated with their course- or lab-based accommodations, particularly if they have not chosen to disclose to this group. Furthermore, some students may view overt accommodations such as technological solutions or technical assistance as "crutches" or as conferring on them an unfair advantage over their peers. Students may worry that accepting accommodations may be considered by some as cheating and therefore choose not to pursue this level of support. While students should never be placed in a position of being forced to accept accommodations or assistance because of their disability, and ultimately, it is the student's choice to engage in the dialog around any accommodation need, it's important to remember that the essential requirements framework requires the student to demonstrate the essential competencies of his or her field. A student with a disability who chooses to eschew the appropriate accommodation—human or technological—and who is as a result unable to demonstrate that he or she can fulfill the requirements of the discipline, program, or course is in a significant bind. As with any other student in those circumstances, faculty would be in their rights to fail or censure them.

This is an important point—while a disability may be invisible to peers, and may be kept confidential by the disability services staff and faculty, an accommodation aid may be a very visible symbol of a student's disability. After all, it is hard to hide something as large as a CCTV. While this may be a concern for some students, anecdotal evidence suggests that some faculty members at the graduate level are equally as concerned about setting up a situation in their research group where not everyone is treated "the same." Unfortunately, these are not easy discussions or scenarios to resolve, as there are significant ethical conundrums wrapped up in this conversation students and faculty must themselves wrestle with: Sameness versus fairness, and the issue of special treatment versus leveling the playing field and providing equal opportunities (Higbee, 2008). Peer support for the student, and for the faculty, as well as having the right mentorship supports, are most helpful in discussing and evolving solutions to these concerns. There is no one answer or approach to situations like this. Each case depends on a number of variables including nontechnological variables such as personalities, lab culture, etc. What might work well in one research group for one student may not be applicable for another student in another scenario. Each situation warrants individual assessment, exploration, and consideration when it comes to implementing accommodations.

Engineering custom technology solutions: the need for a knowledge base

One potential significant side effect of the "trailblazer" phenomenon—the perspective that a student with a disability is the only one in his or her field within the experience of the faculty, accommodation specialist, and student themselves—is that custom technology solutions are trialed in isolation, without recognition of others' prior experience that may be beneficial (both positive and negative experiences—i.e., solutions that worked, and those that didn't). This may lead to "reinventing the wheel" in identifying or developing potential technological accommodation aids for students with disabilities in the sciences. Furthermore, it has the unintended consequence of inhibiting or stifling innovation in the field. Unless previous solutions and avenues of exploration are known, every student, accommodation specialist, and faculty member engaged in this process will start from the ground up.

Anecdotal evidence suggests that there are a lot of potential creative ideas and real solutions that may be found within the community—indeed, the application of student creativity and engagement of others with interdisciplinary collaborative spirit to the cause is quite beneficial, but currently managed only in an ad hoc and largely undocumented manner. However, the development of a "knowledge base" within the community, where appropriate adaptations of mainstream technologies can be captured, documented, and searched, in the context of both field of study or discipline and disability type, is needed. Such a resource would help foster a new sense of innovation, accomplishment, and possibility among students with disabilities in the sciences, their educators, and accommodation specialists, at all levels of education (Higbee, 2008).

Conclusion

In this chapter, we reviewed several case studies for the implementation of accommodations using mainstream and off-the-shelf technology. We also highlighted the potential uses of smartphones and tablets in science lab and fieldwork settings, and discussed the importance of universal design principles in purchasing lab equipment. Adapting mainstream and off-the-shelf technology when accommodating a student with a disability in the lab setting is thus a viable solution and may address many of the learning needs of the student, as well improve the universal accessibility of the lab environment as a whole. Finally, we discussed a strategy that can be used to ensure the most appropriate technological accommodation is designed and implemented for the student.

Assistive technology

Chapter Outline

Introduction   232
What is assistive technology?   233
Barriers to accessing assistive technology   234
Assistive technology in the laboratory setting   235
Meeting the challenges   236
Conclusion   236

My favourite piece of technology is a spyglass.

This handy little telescope – a monocular, really – has been my trusted technological companion since I was in high school. It let me read the blackboard and overhead – and, later, everyone's favorite, the PowerPoint presentations that have become so ubiquitous in scientific presentations, from lectures, to departmental seminars, and at conferences from breakout sessions to plenaries. It lets me engage with the material in a way that I would not be able to otherwise.

Throughout my scientific career, I've had many other pieces of technology – larger, more complex, flashier devices. Definitely more expensive ones. But some of those are, or were, incredibly specialized in their application. It's hard to take a closed-circuit television into a classroom or a seminar, for example – it's not impossible, and other students have had the need to make it happen, but they aren't the most portable thing. I've definitely taken my laptop into those settings, and that definitely is a very versatile tool, but it doesn't help me interact with the information flow in quite the same way.

I've even tried different kinds of telescopes – binoculars, devices mounted on my glasses, hands-free options, handhelds with different focal lengths and magnifications. This one model, though, has always worked the best, and has been the one I've returned to time and again over almost twenty-five years. I've had handheld magnifiers for different applications – again, some more fancy than others – but nothing I've felt as visceral a connection to as my spyglass.

For different people, the technological favourite will be different. For some, it might be a Braille note taker. For others, their CCTV, or their laptop, or the digital tape recorder they use to record lectures and seminars. Or their FM system and microphone. Or an ergonomic chair. It will be whatever they are using that helps them engage with their learning as a scientist the most. And whatever they have the most visceral reaction to doing without in the settings where they need it.

The funny thing is that some of these pieces of assistive technology don't seem...well, very technological.

A magnifier?

A telescope?

A chair?

There is a perception, I think, that says that "assistive technology" means "complicated" or "expensive" or even "big" – assuming, that is, that people understand the term at all. Of course, that's not true: assistive technology – or AT, for short – is merely any piece of technology that helps us as persons with disabilities succeed at whatever it is we are trying to do (in this case, learn and participate in the sciences). Just because the field is technically complex doesn't always mean that we need complex things in order to engage in the learning and work environments.

It's very easy to get distracted by the more complex pieces of technology – people try to sell us on them all the time, and, from a naïve perspective, if we haven't fully thought through the problem, it might make sense to go that route. To be sure, those more complex technologies are needed in some instances – my retrofitting a CCTV to accommodate animal dissections would be one example of this – but I've found that taking a more holistic approach of finding the right technological solution to the problem at hand is the way to go.

Sometimes the simplest AT solutions are the best.

Introduction

Today, technology is crucial in all educational, employment, and recreational activities. Computer access has the potential to help people with disabilities complete coursework independently, participate in class discussions, communicate with peers and mentors, access distance-learning courses, participate in high-tech careers, and lead independent lives (Burgstahler & Cronheim, 2001). Students with disabilities benefit from the use of technology in the same way as their able-bodied counterparts. Just as every disability is unique, so too is each individual's need for and use of assistive technology (AT).

Although the benefits of technology may be even greater for people with disabilities than for those without disabilities, individuals with disabilities are less likely to own a computer or to use the Internet (Kaye, 2001). Even for those who can operate a computer, the design of many webpages, instructional software programs, and other electronic and information technologies creates access barriers (Lane & Mistrett, 1996). For example, webpages that do not include text alternatives that can be read by speech and Braille output systems limit information access by a student who is blind; the content of multimedia resources that do not have captions is inaccessible to a viewer who is deaf; and scientific equipment that cannot be operated from a seated position is inaccessible to a student who uses a wheelchair.

Many technology products are designed in such a way that they are inaccessible to people with some types of disabilities. For example, a person with a visual impairment may not be able to interpret telephone use instructions if he or she is presented with instructions only in a visual format, and a person who is deaf cannot access content of a software program that is only presented aurally. Too often, even those individuals with disabilities who have AT, a computer, and an Internet connection still cannot make full use of technology because of the inaccessible features of hardware or software. In contrast, when universal design principles are employed as technology is created or updated, the resulting products are fully accessible to a broad audience, including AT users.

What is assistive technology?

According to Mohler (2012), AT is "any item, piece of equipment or product system, whether acquired commercially off-the-shelf, modified or customized, that is used to increase, maintain or improve functional capabilities of individuals with disabilities." AT can be simple or complex. Examples of low-tech tools for students with disabilities might include enlarged text or raised line paper, while high-tech tools may encompass digital tools that "read" to the student, connect to a Braille display, or even incorporate GPS. AT can be broadly classified into three distinct categories:

- low-tech AT
- high-tech AT
- software solutions

Low-tech AT includes "a variety of simple tools and commercially available devices and solutions that are easy to use, inexpensive, and require little training" (Lane & Mistrett, 1996). Low-tech AT solutions may include but are not limited to:

- magnification devices
- facilitated communication keyboards
- Ergonomic solutions for office/workstation set up
- wrist braces
- large-print calculators
- monoculars and small telescopes for distance viewing
- tape recorders
- accessible landline phones
- modified computer mice

High-tech AT, by contrast, is more expensive, requires more training, and is more complex in terms of electronics and to operate. High-tech AT solutions may include but are not limited to:

- laptops
- closed-circuit televisions (CCTV)
- Braille displays

- iPads, iPhones, and other tablets and mobile devices
- talking scientific calculators
- infrared and FM systems
- talking GPS

Software solutions are a specific category of AT that enables a user to interact with technology and gather data about the surrounding environment. Software solutions may include but are not limited to (Raskind & Higgins, 1998):

- screen-reading software
- screen-magnification software
- text-to-speech synthesizers
- optical character recognition (OCR) software
- speech recognition software
- word-processing, spell-checking, and proofreading solutions
- outlining/brainstorming software tools

Barriers to accessing assistive technology

Computers, mainstream and AT, and web-based resources can bridge the communication and accessibility gaps for people with disabilities. However, in order for this gap to be bridged, it is critical that the interface between the student's AT and the relevant mainstream platform be accessible. Electronic communications provide options for independent access to people and resources. Computer and network access can increase levels of independence and have a positive impact on the academic progress and career success of individuals with disabilities. Unfortunately, many individuals with disabilities and people in their primary support networks are unaware of the tremendous contributions technological innovations can make to the lives of individuals with disabilities (Hewlett & Burnett, 2006). Students with disabilities are not guaranteed access to computing and networking technology in science, technology, engineering, and mathematics (STEM) programs, particularly in resource-poor regions and countries. Likewise, lab facilities and electronic resources are often designed in such a way as to be inaccessible to students with disabilities.

Those who wish to pursue STEM fields need access to publications in these fields, yet STEM publications are not always readily available in alternate formats. Making them available in an accessible electronic format is desirable, but some barriers still exist in making mathematical and scientific symbols as well as graphical images accessible to those who are blind. Universal access to publications will require the creation of new products, as well as promotion of the use of existing methods. Webmasters also need to apply standards, such as those used by the US government, in order to take steps toward making their resources accessible to individuals with disabilities, including those who are blind and use text-to-speech technology.

In order for students with disabilities to pursue postsecondary education, or careers in STEM, they must have access to the high-tech tools available to their nondisabled peers. These include computers, websites, Internet-based distance-learning courses, instructional software, and scientific equipment. Achieving this goal requires that (a) appropriate AT be readily available, and (b) barriers to electronic tools and resources be eliminated.

For example, it is important that students who are blind have access to speech and/ or Braille output devices. But access to this AT is not enough. In order for students to benefit fully from the use of this software, the applications used in the laboratory and classroom setting must be fully accessible and compliant with AT. For example, a student who is blind and has been provided with screen-reading software on his or her computer in the laboratory will not be able to make full use of that screenreader's function if the web-based course platform is not accessible with screenreaders and cannot be navigated using keyboard commands. There are also issues to consider around the expense of most AT, and the lack of funding available to provide students with disabilities with this type of technology (Lane & Mistrett, 1996). Additionally, it is critical for students with disabilities to be knowledgeable about how and where to acquire the technology they require in the laboratory or classroom. It may be the student's responsibility to educate faculty, teaching assistants, and staff on the technology he or she requires and its use (see chapter 7: Student as ACTor — recognizing the importance of advocacy, communication and trailblazing to student success in STEM).

Despite several developments in enhancing access to math and science content, there continues to be significant challenges in truly enabling students with disabilities to take advantage of the STEM resources on an equal basis. This is primarily due to two problems:

- Despite the availability of several products and techniques, there is very little knowledge about these tools and strategies, even within the disability sector.
- Several tools are pretty costly (which is especially true for students in developing countries). Yet other tools cause ambiguities when interpreting the results. For instance, OCR tools do not provide 100% accuracy. Similarly, when complex equations are read with MathPlayer, there tends to be a lot of confusion in analyzing the equation (Isaacson, Srinivasan & Lloyd, 2010).

Assistive technology in the laboratory setting

Students and researchers with physical limitations are hindered from independently performing many hands-on tasks in the laboratory setting (Sukhai et al., 2014). Light microscopy (LM) is one of the most essential laboratory techniques performed by students in STEM courses, including biology, natural sciences, and food sciences. The ability to independently operate a light microscope provides students an active learning experience that is necessary for enhanced recall of LM procedures and a more in-depth understanding of many biological sciences concepts. Knowing how to independently operate a light microscope is important for students wishing to pursue graduate research in some fields in the biological sciences. To enable students to independently operate a light microscope, an automated microscope workstation called Access Scope was developed for people with upper-limb impairments and visual impairments. Access Scope allows the user to operate all features of a light microscope including focusing, changing objectives, illumination settings, and exposure rates and loading or changing slides without human assistance. The Access Scope enables students with some usable vision to independently perform LM and related research tasks such as image analysis.

Another AT tool that permits students who are blind or who have low vision to independently navigate the research and lab environment is the Talking LabQuest device. This tool speaks specific data points out loud to the student. Its small size minimizes the space required on the lab bench, and allows it to be used in the field and in informal learning environments. This device can be hooked up to a PC to transfer data gathered while in the laboratory.

Meeting the challenges

Options that can be considered to meet access challenges in the laboratory include:

- Stakeholders should have access to training so they can design and select accessible facilities, utilize computers and software, purchase appropriate AT, and ensure students with disabilities use technology for their maximum benefit as they pursue academics, careers, and self-determined lives.
- Policies and procedures should be established at all academic levels to ensure universal accessibility is considered when electronic and information technology is procured.
- Policies, procedures, training, and support should be established at all educational levels to ensure that webpage, library resource, and distance-learning program developers make their electronic resources accessible to everyone.
- Interagency collaboration on planning, funding, selecting, and supporting AT should be fostered to ensure continuous technology access and support as students with disabilities transition through academic levels and to employment.
- Students with disabilities should be included at all stages of technology selection, support, and use, so they learn to self-advocate for their needs for AT in the classroom and workplace.
- Students with disabilities at high school and college levels should participate in internships and other work-based learning experiences where they can practice using technology in work settings.

Legislators and policymakers should disseminate information about current laws, policies, and resources that are universally designed to meet the needs of various stakeholders. They should also identify and correct inconsistencies and gaps in legislation and policies regarding the selection, funding, and support of technology for people with disabilities.

Conclusion

Access to electronic and information technology has the potential to promote positive postsecondary academic and career outcomes for students with disabilities. However, this can only be achieved through greater awareness of the use and applicability of AT, and through ongoing compliance to web and course platforms. However, this potential will not be realized unless stakeholders (a) become more knowledgeable about appropriate uses of technology, (b) secure funding, and (c) work together to maximize the independence, participation, and productivity of students with disabilities as they transition to college, careers, and self-determined lives. Ultimately, ensuring all of the educational and employment opportunities that technology provides are accessible to everyone will strengthen our economy and contribute to the creation of a level playing field. Awareness of the various types of AT and their applicability to STEM disciplines is the key to empowering students with disabilities and helping them pursue STEM fields, which was hitherto considered impossible.

On the international stage, in countries and regions where access to technology solutions in the educational context as well as the research infrastructure are limited it is important to recognize the problems of scope introduced here and in the previous chapter. However, there is significant room for interagency collaboration and partnership around the identification and provision of appropriate solutions. Cooperation among United Nations agencies, nongovernmental organizations in education and development, advocacy bodies, and groups of educators, accommodation specialists, and students is encouraged to resolve this challenge. Specifically, educators, and accommodation specialists, through their own networks, may serve as champions for their students in the sciences, and in so doing, can identify the right people and groups to initiate a dialog with.

Accessible formats in science and technology disciplines

Chapter Outline

Introduction 240
What are print disabilities? 240
What are accessible formats? 241
Accessible format materials and technologies 241
Challenges with accessing accessible formats in the classroom setting 242
Accessing accessible format materials in the laboratory setting 243
Tips for making accessible formats accessible in the classroom and laboratory 244
Accessibility and online learning environments 245
The Marrakesh Treaty 246
Conclusion 247

Information flow is important to me. We've established many times over how much I love science, and as my career has evolved, my primary approach to learning has been to access information (either intended for print, or made to look like print) in different ways. Science is very technical by its very nature, and most scientific print material is simply not presented in a usable form for me. The font is super tiny and very dense. Graphs are small, busy and often unclear on the page. Contrast in images and photographs may be too low for me to resolve clearly.

When I started my Master's, in order to read articles published in scientific journals, I used to go to the library, find the journal in the stacks – a chore in and of itself, given poor lighting, narrow aisles, and small print on book or journal spines – photocopy the article I wanted, and then get it enlarged on copiers that could handle the paper size and enlargement options I required. I used to have filing cabinets full of these articles.

Today, I'd never do that – I would simply carry out an electronic search for the article I wanted, and since most everything is online, I can just download it and store it on my computer. Even better, I can blow things up as big as I need to on a sufficiently-sized computer monitor – or, with the right format file and screenreader software, I could get the article text read to me. Some readers of this book might be employing that method right now to read these words.

I read best with large print. Although I learnt Braille, I didn't continually practice, and so my knowledge of that language became rusty and slipped away over time. My reading method works, for the most part, with journal articles – it doesn't work for scientific textbooks or reference materials, though, as they may not be online or in e-book formats. It also doesn't work in more spontaneous settings, or settings where I can't receive materials to review in electronic format.

Scientific conferences are especially problematic for me, in terms of being able to read materials in an accessible way – usually because the volume of material is in such tiny print, and it's not always practical to track it down online when one is rushing between simultaneous sessions or between posters in a crowded convention centre. That's where lots of advance planning becomes important, but that can sometimes rob one of the spontaneity of saying "Oh, that sounds interesting! I'll go."

I've learned that it's good, then, to have multiple possible backup approaches to reading materials in accessible format – sometimes, large print will work; other times, e-text; other times still, a human reader. As with so many things in the sciences, because of the diversity of the potential learning environments, the extreme specialization of the material and its technical nature, as well as the general challenge around print sizes, it is good to be flexible, and to recognize that I may not have an ideal solution worked out well in advance.

It's worth it, though, to make sure that I can continue to engage with the field I love.

Introduction

In this chapter, we discuss alternative formats for reading print material such as audio, Braille, large print, and electronic text. We describe the different types of accessible formats available, and how each can be used in the classroom or laboratory setting. We also present creative, low-tech solutions to creating materials in accessible formats, and we explore some common barriers faced by students with disabilities when using online course platforms.

What are print disabilities?

Standard printed publications are not accessible to people with visual impairments, or who are blind. Many other types of readers also cannot use printed books, newspapers, and magazines—including those with dyslexia and other types of learning disabilities as well as individuals with motor disabilities or manual dexterity limitations who cannot hold or turn pages in a book. Collectively, these types of disabilities are called "print disabilities."

What are accessible formats?

Accessible formats are other ways of publishing information aside from standard print that still provide the same information, just presented differently (Rowlett & Rowlett, 2009). Some of these formats can be used by everyone, while others are designed to address the specific needs of someone with a disability. For example, use of Braille materials is a specific modification for those who are blind or who have low vision, while e-text is used by everyone.

It is important to note that accessible formats are required for a variety of learners, in a variety of contexts. For instance, a student with a disability may require accessible format material in his or her classroom, laboratory, or fieldwork placement and may require different types of accessible format materials, depending on the context. As the nature of the work conducted in these settings differs, so too will the type of format required. For example, a visually impaired student may need large print or Braille equipment labels in the lab, but e-text of scientific articles would be sufficient when he or she is doing required readings for class or research. Disability can also have an impact on the type of format needed; students with visual disabilities require formats that translate print into audio or tactile graphics, whereas a student with a hearing impairment requires audio materials to be translated into visual, written words.

Accessible format materials and technologies

The following is a list of some of the more common forms of accessible format materials or technologies and the specific groups they may benefit the most.

Large Print: Large print is most helpful to people who have low vision. Large-print materials should be prepared with a font (print) size that is 16–20 points or larger. The size of the font should be calibrated to the visual acuity of the reader.

Braille: This is a language for people who are blind or deafblind. Braille is a tactile system of raised dots representing letters or a combination of letters of the alphabet. Braille can be produced using Braille transcription software, or by hand. Using various specialized devices (e.g., a Braille notetaker with a refreshable display), people who are blind or deafblind can also take notes in Braille; this can also be done manually using a stylus. Specialized versions of Braille have been created for technical disciplines such as mathematics and the sciences, and can be learned and used by students in these contexts.

e-text: E-text, simply put, is a word-processing file containing the text of a book or other printed document. E-text can be read in large print on the computer screen, or by a screenreader. E-text documents can also be read on devices such as tablets, and can be created in a variety of ways—if the original material is presented in print, it can be scanned into software that recognizes and converts the printed material into readable audio formats that can be read with a screenreader.

Screenreaders (text-to-speech engines): This software converts text that is displayed on a computer monitor to voice (using a speech synthesizer). This technology is most appropriate for people with a vision, intellectual or developmental, or learning disability who are unable to read print, and when there is a version of the document to be read available on a computer.

Tactile graphics: Many technical concepts in the sciences are illustrated visually, in ways inaccessible to people who have visual impairments and to everyone who is totally blind. Tactile graphics are line drawings specifically created so that they can be read and interpreted solely by touch, and if necessary, can be annotated in Braille. Several groups and companies have created tactile graphics for a variety of science concepts in education. Related to the concept of tactic graphics is 3D printed material for the blind or partially sighted. 3D printing enables a more intimate interaction with the object, and may even permit for disassembly and reassembly to aid in learning.

Audio format: This is an accessible format for people who are unable to read print in which a document is read and recorded for later playback. If material is available electronically, screenreader software performs the same function.

Captioning: Captioning translates the audio portion of a video presentation using subtitles or captions, which usually appear on the bottom of the screen. Captioning may be closed or open. Closed captions can only be seen on a television screen that has a device called a closed caption decoder. Open captions are "burned on" a video and appear whenever the video is shown. Captioning makes television programs, films, and other visual media with sound accessible to people who are deaf or hard of hearing.

Windowing: Windowing lets people who are deaf read by means of an interpreter who explains using sign language what other people are hearing during a video presentation or broadcast. The interpreter appears in a corner or "window" in the screen translating spoken language to sign language. Windowing may include open or closed captioning.

Descriptive video service (DVS): DVS provides descriptive narration of key visual elements—the action, characters, locations, costumes, and sets—without interfering with dialog or sound effects. This makes television programs, films, home videos, and other visual media accessible for people with vision disabilities.

Challenges with accessing accessible formats in the classroom setting

There are various challenges associated with accessing technical information in accessible formats. Accessing equations via Braille or audio formats is challenging because these formats are by nature linear, while a printed equation gives more of an overview (Rowlett & Rowlett, 2009). Similarly, with large or magnified print, it is difficult for a student to see a whole equation all at once, so the student has to rely on memory to recall the information presented in the equation, as the whole equation is not visually accessible to them. This places the student at a great disadvantage, as he or she cannot access the information presented in the same manner as his or her peers (Beal & Shaw, 2008). Further difficulties with large-print versions of resources include where to break equations that do not fit on a single page (Rowlett, 2008; Rowlett & Rowlett 2012). This can also be a problem in electronic formats (such as

Word, LaTeX, and MathML), which have limited automatic line breaking of equations. This means if some notes are made accessible at one size and then need to be altered to a different size the whole document may need re-typesetting (Cliffe, 2009).

Using audio—either via technology or using a human reader—can be ambiguous. For example, knowing where to place punctuation when reading a sentence or phrase can completely alter the meaning of the text. One popular example comes from the book *Eats, Shoots, and Leaves*. Placement of the commas, or lack thereof, can change the intended message of the text. Accuracy in reading the punctuation reflected in the text is critical when accommodating learners with visual difficulties.

The difficulty of producing technical material in accessible formats creates challenges for learners in the math and sciences not only for accessing key materials such as lecture notes, but also to carry out further reading around the subject (Rowlett & Rowlett, 2009). As conversion to accessible formats is time consuming and costly, blind/partially sighted learners usually only have access to materials such as core texts, which means they often do not have access to the same materials as their peers.

Initiatives have been undertaken to improve access to science, technology, engineering, and mathematics (STEM) materials. For example, the program lead by The University of Bath that developed comprehensive guidance for higher education staff on the subject of producing accessible learning resources in mathematics.

Accessing accessible format materials in the laboratory setting

The major challenge facing students who require materials in an accessible format is the overwhelming amount of visual material to which they are continually exposed, such as textbooks, class outlines, class schedules, chalkboards, writing, models, images and other graphic materials, etc. The use of films, videos, computers, and television programs also adds to the volume of printed material to which the students have only limited access. To assist in teaching a student with a visual impairment, unique and individual strategies should be employed that are based on that student's particular disability and specific learning needs.

The following is a sampling of some of the many accessible format materials that are available to students with disabilities in the science laboratory:

- Verbal descriptions of demonstrations and visual aids
- Braille text and raised-line images
- Braille or tactile ruler, compass, angles, protractor
- Braille equipment labels, notches, staples, fabric paint, and Braille at regular increments on tactile ruler, glassware, syringe, beam balance, stove, other science equipment
- Different textures (e.g., sandpaper) to label areas on items
- Large-print, high-contrast instructions and illustrations
- Raised-line drawings or tactile models for illustrations
- Large-print laboratory signs and equipment labels
- Large-print calculator (Burgstahler, 2012)

Tips for making accessible formats accessible in the classroom and laboratory

As previously noted, clarity in describing written content is essential for students with disabilities. Indeed, being specific when referencing written content on a smart board or PowerPoint will not only benefit those students with disabilities in your class or lab, but will assist all students in reinforcing the material being taught. For example, say "benzene," not "this compound." Say "from 20°C to 40°C" instead of "from this temperature to that." State "in the reaction of benzene with bromine" rather than "in this reaction."

High-contrast smart boards are available that generate printed copies of board contents. Again, instructors can facilitate learning among all students by providing copies of the presentation, including smart board-generated material and visuals.

Projection systems and computerized presentation programs like PowerPoint allow users to manipulate brightness, contrast, fonts, graphics, and colors, which may be helpful for students with low vision. It is important to consult students to determine the best format. They may benefit from the use of colored acetate overlays, e.g., to improve contrast of projection system presentations. Digital formats also make it easy to produce printed copies of presentations or to provide students with an accessible format to meet their needs.

Many instructors already prepare for their classes in ways that help ensure access for students who are blind or vision impaired. Some use standard techniques for effective teaching such as preparation and frequent updating of a class syllabus, making course materials available to students in digital formats, posting course materials to a web page, and accepting assignments submitted as e-mail attachments. These approaches also are becoming more common at the K–12 level. Some teachers may prepare for all courses with the assumption that a student with a disability will enroll. This approach can save time and effort when such a student does join the class. It also offers greater accessibility to course materials and facilitates learning for all students.

Given the time it takes to convert textbooks to an accessible format, it is important to select textbooks and other required reading material to allow time for this process. When selecting a new textbook, consider texts that are available in both standard print and digital or recorded format. Access to a digital format greatly reduces the time and expense of converting a text into electronic or other accessible formats. Alternatives to the standard textbook format may also be important for students with certain learning disabilities. Check with the disability services office about the time needed to convert overheads, slides, and other printed materials into accessible formats. Be sure to send copies of the material to the DSS office within that time frame. Materials should be available to the student in an appropriate accessible format at the same time they are distributed in print to other students.

Prepare all classroom handouts as digital files that can be easily converted into accessible formats. Digital format also is ideal for students with visual disabilities who use computerized assistive technology such as magnification or screenreader software. When creating lecture notes, provide notes in an easily accessed digital format such as text or PDF. Graphics and images should be fully captioned, with detailed descriptions, and should be integrated with the rest of the material.

Accessibility and online learning environments

E-learning (commonly referred to as distance learning, online learning, or web-based training) involves instructors educating students through a variety of digital media and offering an accessible to a one-size-fits-all approach to learning in a traditional set time/set location classroom environment. Accessibility of online learning resources is crucial to ensure everyone can benefit from the unique opportunities afforded to students using online learning platforms.

These web-based technologies deliver audio, images, text, and video content to students enrolled in an online course or training program. E-learning can be coordinated online or in the classroom. Many institutions also offer hybrid or blended courses that give students the option to attend class either online or on campus. E-learning can be instructor-led and happen in real-time (synchronous learning) or allow students greater flexibility to complete courses at their own pace.

Online learning is an ever-changing emerging market; technologies are being developed and replaced at a rapid pace. It is imperative that these technologies don't leave people with disabilities behind. Online courses and training programs are created using a platform referred to as a learning management system (LMS). There are many different systems on the market today; two common choices are Moodle (an open-source platform) and Blackboard (a proprietary platform). Regardless of what platform an organization uses, it should be accessible for students with disabilities. If portions of the course are inaccessible, students may be put at a disadvantage.

Designing accessible courses means all students have a better experience, whether they have a disability or not; if accommodations are needed, very little extra work goes into providing them. Any time a course includes a video, consider including transcripts, which are accessible for screenreader users, work at low bandwidths, are printable, and are preferred by some students who would rather read the content than watch a video. Providing the content in multiple formats accommodates students with disabilities, students on various devices and platforms, and students with varying learning preferences.

STEM courses pose unique accessibility challenges. Math symbols and scientific notation are typically displayed as images in lesson hand-outs and PowerPoint slides. By themselves these images are not accessible (because screenreaders can't interpret graphics) and providing accurate accessible text descriptions can be challenging. Special code for math content was recently developed by the W3C Math Working Group called MathML, which allows the content to be displayed in a screenreader-friendly way. As an added benefit, the coded content is easily translatable into a number of languages, including Braille.

The following is a checklist of best practices instructors and course designers can use to ensure their online course platforms are fully accessible

- Use text descriptions for image-based or visual content.
- Provide a transcript for audio content.
- Include captions and, where feasible, audio description for video content.
- Consider providing video, audio, or graphics that express equivalent meaning for text content.

- Use colors that contrast well.
- Use actual text rather than images with text embedded in them.
- Be sure you can follow the cursor easily when navigating with the tab key.
- Do not use color alone to represent meaning.

(List adapted from Council of Ontario Universities, Accessibility in E-Learning, 2016.)

The Marrakesh Treaty

The Marrakesh Treaty was adopted on June 27, 2013 in Marrakesh, Morocco, by the World Intellectual Property Organization (WIPO). The goal of the treaty is to increase the amount of accessible format text available to those with print disabilities. Despite the ability to convert print books into accessible formats like Braille, audio, and digital copies, over 95% of published works are unavailable to people with print disabilities. A work printed in English may have already been converted into a format that is accessible to those overseas who cannot read standard print, but because copies are not exchanged across borders, domestic entities might need to make a duplicate copy or just might deny access altogether by failing to reproduce the work. Intellectual property (IP) legislation is, by nature, territorial. This means that the rights, obligations, and limitations expressed in copyright law apply exclusively to the country where the law is enacted. Thus, by 2008, a relatively large number of countries had some provision in their IP legislation for the production and distribution of accessible format copies, but this exception could only be used at a national level (Sullivan, 2006). An uncoordinated legal approach prevents the cross-border exchange of accessible books. Two-thirds of the world's nations do not have domestic copyright laws that permit making copies for those who are print-disabled, limiting the number of works available in an accessible format. Moreover, many countries consider distribution of accessible copies an infringement, and even among nations that permit distribution, limitations vary. Instead of exchanging books across borders, works are needlessly duplicated, and circulation is significantly limited. Although the Marrakesh Treaty will greatly improve access to printed material for students with print disabilities, course packs, lab manuals, and academic journals remain outside the scope of it. The accessibility of course packs and lab manuals remains the responsibility of individual faculty members, departments, and institutions, and journal publishers are responsible for their specific articles. Students will still need to rely on requesting that print materials be converted to an accessible format by the student's postsecondary institution. Finally, heavily technical material represented as graphics, equations, and line drawings does not fall under the purview of the Marrakesh Treaty. Thus ensuring the accessibility of these materials continues to happen on an ad hoc basis, and remains the responsibility of individual faculty.

The Marrakesh Treaty is a landmark accomplishment for the production of copyrighted works in accessible formats, and the sharing of such materials across international boundaries. Notwithstanding the technical challenges of producing accessible materials in the sciences, creating science textbooks and related works in an accessible format is expected to, over time, mitigate a substantial barrier to accessing information that exists internationally for students with disabilities studying in the sciences.

Conclusion

Accessible format materials are used by students with print disabilities. Depending on the nature of the student's disability, and/or the specific course, this will influence the format of the material being produced. There are various issues with accessing technical information in accessible formats. These can range from translating printed equations into a tactile or electronic format, to translating labels and other lab materials into a tactile or large print format, to converting text to be readable with a screenreader.

Simulation learning

Chapter Outline

Introduction   250
Virtual learning in the sciences   251
Simulations and learning styles   252
Considerations when applying simulation learning to the sciences   252
Simulation learning and accessibility   253
Simulation learning as accommodation   253
Simulation learning as a course/program component   254
Simulation learning in postsecondary education   254
Conclusion   255

It's amazing how far technology can advance in twenty years. When I began my undergraduate education, the only simulations available to me were observation of hands-on demonstrations in the lab. If something was too dangerous for the class to do, the lab instructor did it, and we all watched. Now, of course, we can simulate many such experiments using computer programs, or even YouTube tutorials – often with appropriate sound effects and visuals.

Today, we can learn dissection and surgery virtually – no messy blood. Or, in my case, no need to worry about how poor my surgical skills really are, because of my hand–eye coordination. Of course, there was no way to simulate the results of the animal dissections we had to do during my doctoral program – no way to use a computer program to predict what we would have found in those mice. That's biology for you – messy, complicated, and, some would say, irreducibly complex.

Had I stayed in physics, things might've been different – I was a reasonably good computer programmer back in the day (even though I didn't like the subject). I'd have probably had to develop my own simulations to model my research question. I had wanted to be an astronomer – perhaps I'd have been on one of the teams to have modeled the existence of exoplanets from stellar observations. Perhaps I'd have developed educational simulations in astronomy and physics for my students to learn from. I like to think I'd have done them in an accessible manner, compliant with the appropriate web and software development guidelines.

Simulation learning today is such a burgeoning field – indeed, we even call it different things. "Game play," for example. I remember learning projectile motion by playing a computer game—and that was before I met the concept in high school physics.

It's easy for educators to suggest simulation learning as a primary fallback for a student with a disability in the sciences. "You can't be in the lab – play this game instead!" We have to be careful about that though – tempting as it might be to say "yes" we need to make sure that the simulation achieves the same goals as the exercise we're missing, and the interactions we lose out on by doing the simulation. We also need to make sure the simulation will work for us – our accessibility requirements as well as our learning styles.

For someone like me, who learns by hearing, reading, journaling, and teaching, a very visual simulation would do no good. Indeed, a dissection simulation, as tempting as it might be, is almost as useful to me as actually dissecting an animal – which is to say, not in the slightest. They are both very sight-oriented activities, and if I had the dexterity, I'd frankly prefer the hands-on surgery. The ultimate test of a simulation's worth is if it adequately prepares a student for the real-life counterparts down the line once they have moved on to the next stage of their careers.

For my students, were I in a position to determine whether a simulation would be appropriate as an accommodation, I think I would try consulting with them. They know their disability best, and are best positioned to tell me what benefits a simulation might have to offer. If I were trying a simulation as a classroom activity, I'd want to be careful about my choice of simulation, and want to make sure it's as accessible as possible. Much of that is because I have lived experience, and I'd try to put myself in the position of my students with disabilities. I sometimes think that's perhaps the best type of simulation learning out there.

Introduction

Simulation learning can be a valuable tool to support the learning of students with disabilities in the sciences. Specifically, simulation learning is a generic term that refers to an artificial representation of a real-world process to achieve educational goals through experiential learning (Abdulmohsen, 2010). Simulation-based education is defined as any educational activity that utilizes simulation aides to replicate real-world scenarios.

Simulation learning is traditionally thought of as *virtual learning* or *computer-aided simulation*: a student sits in front of a computer, or computer-driven simulator, and watches or participates in an online or virtual experiment or demonstration. One such example of a simulation learning experiment is the digital dissection of a frog.

Simulation learning can come in several additional flavors. For example, *assisted learning* is a scenario where an assistant sits with a student, and the student provides direction on the steps of the experiment (e.g., a laboratory or technical assistant; see chapter 19: Mainstream technology as accessibility solutions in the science lab, for more details). In the context of *mechanical assistance*, a machine is used as a physical surrogate, performing defined tasks in lieu of assistance provided by another person (e.g., robotics in a science lab environment). The technologies used in mechanical assistance are not specifically designed for people with disabilities, but instead are mainstream technologies that can be adapted for specific applications in the learning environment.

Educators can also use simulation learning in the classroom for all students, particularly in settings where the topic being studied cannot be modeled in a traditional teaching laboratory environment. Indeed, a field of research has grown up over the last decade or so around the use of computer games and other high-quality simulations as teaching tools in the sciences at various levels of education.

Thus, in thinking about simulation learning, we must consider two scenarios: (1) Situations where simulation learning will benefit a student with a disability specifically, because other accommodation methods (e.g., technical assistance) are not appropriate or feasible; and, (2) situations where simulation learning is applied to all students, in which case accessibility considerations of the choice of simulation must be taken into account for students with disabilities in the class. In this chapter, we will review the application of both scenarios for simulation learning to students with disabilities in the sciences.

Virtual learning in the sciences

There are three broad categories of virtual learning approaches that can be employed. The first of these, like the virtual dissection of the frog example from above, simulates a process where there is only one "right approach" or a single group of right approaches—in these types of scenarios, an incorrect step suggested by the learner will lead to an error being recorded or flagged during the simulation. The second type of virtual learning approach is a simulation designed to converge on any possible failure mode (the "no-win scenario," like *Star Trek*'s Kobayashi Maru), irrespective of learner input. The third type of virtual learning approach involves a simulation that is responsive to the starting conditions and additional inputs of the learner. This last kind of simulation learning is often found in the physical sciences, where system behaviors are better understood and more easily modeled mathematically.

Virtual learning approaches in the sciences vary depending on the field and learning objective of the activity. Particularly in the context of the physical and computational sciences, virtual learning is employed to model situations not reproducible in a laboratory environment, due to complexity, practicality, scale, and cost. For example, the test required to assess various permutations of a given variable can be cheap and quick, but the sheer number of trials might render the experiment impractical to carry out in person, in which case a computer-aided simulation is the solution of choice. On the other hand, simulations in the biological sciences are intended to reproduce or demonstrate real-life processes and procedures.

Finally, virtual learning approaches rely on a well-understood system being modeled or simulated. In new fields, where the mathematics of modeling is not well understood, or in fields where the technology or system is still being trialed, simulation learning approaches are less likely to be successful (i.e., reliable or accurate to the real system being modeled), because the computational modeling required for coding of the simulation would still be in development.

Simulations and learning styles

Although there are multiple ways a person can learn—and different people will have different primary and secondary learning styles, independent of disability—simulation learning approaches are unable to fully replicate the variety of learning styles that are engaged through in-person, hands-on learning. Indeed, simulations are often designed to engage one or two learning styles, at most, and activate a fundamentally different set of learning styles than "hands-on" activities (i.e., in most cases, there is no learned "muscle memory" of the activity gained through practice). Thus students who learn best by physically doing and engaging with the experiment would not learn well through a simulation exercise. There are some exceptions to this rule in complex technical fields (e.g., flight simulators in aeronautical and aerospace engineering), where the learner is able to conduct hands-on simulations of complex tasks (D'Angelo, et al., 2014).

Considerations when applying simulation learning to the sciences

Traditional simulation learning allows the student to acquire skills through the use of technologies to simulate the experiment or exercise (Abdulmohsen, 2010). Simulation tools serve as alternatives to conducting experiments or directly performing required tasks in a laboratory or practical space setting. One key construct that differentiates simulation learning from other forms of learning is that in many cases, the student is "virtually" present, rather than physically present or engaged in the learning environment.

Of the three types of simulation learning, technical assistance provides the opportunity to have the student most closely associated with the task(s) (Allan, 2014). Virtual learning, on the other hand, can be most removed from the laboratory/practical space setting (depending on the exact nature of the simulation), and may offer the least opportunity for direct interaction with the task(s). Technical and mechanical assistance are potentially costly and resource-intensive to implement and maintain. It may be difficult to recruit and retain technical assistants with the appropriate skillset and qualifications. The applicability of mechanical assistance is dependent on readily available technological resources, but since many of these pieces of equipment are expensive to purchase and maintain, they may be outside the scope of many learning environments. Virtual learning simulations of high fidelity, quality, and accessibility may also be too expensive for many educational settings; however, the academic setting does afford opportunities for creative inter- and multidisciplinary collaborations involving the educator, the student, and other collaborators, in order to find effective solutions to some of these problems.

Because simulation learning lacks a direct "hands-on" component, it is important to determine upfront whether simulation learning allows the student to achieve all the required competencies that would be met through a direct physical interaction with the task (see chapter 10: Essential requirements and academic accommodations in the sciences). Without this assessment, it is possible students may miss key steps in the learning process. Relying on virtual learning can be difficult, as the student does not have the opportunity to interact with the required tasks in the same manner as if a laboratory/technical assistant were provided. Where possible, virtual learning should be a secondary option, only used in cases where a laboratory or hands-on experience is not feasible.

Simulation learning and accessibility

In considering the application of simulation learning as an accommodation for a student with a disability, it becomes important to distinguish between the *quality* of the simulation and the *accessibility* of the simulation. Since the simulation is, at its core, a computer program—either run on a computer workstation, or run on a dedicated simulator—the accessibility of the simulation is very dependent on the accessibility of the simulator and the interface between adaptive software and the simulation program on the computer.

The quality of the computer-aided simulation, on the other hand, depends on the validity of the underlying modeling and computer code, and the ability of the simulation to achieve the designated learning objectives. Ultimately, any new simulation designed from the ground up should be designed with the learning objectives, learning styles to be engaged, and universal accessibility in mind—it becomes difficult to retrofit accessibility onto a simulation, but it is also important to not compromise the learning objectives for accessibility either. Ultimately, while the quality of the simulation is not an accessibility issue, accessibility is a quality issue, and should be considered as part of the design process.

Simulation learning as accommodation

With the caveats we have discussed in mind, let us return to the two scenarios in which simulation learning can be applied for students with disabilities. In the first scenario, simulation learning is used in place of another approach as an accommodation, after other forms of accommodation that are more "hands-on" are deemed inappropriate.

As with our discussion around technical assistance in the previous chapter, it is important to evaluate the simulation learning approach against the essential requirements of the course, program, and/or discipline. However, rather than determining whether the simulation learning approach contravenes the requirement to do the work independently, the core issue is whether the simulation learning approach enables the student to meet the core competencies set out for the program, course, or discipline. Specifically, if a range of simulation learning approaches are available, the best choice is dictated by the accessibility of the simulation and the degree of match of the simulation's objectives to those defined by the essential requirements. Often this will require some significant upfront testing of possibilities and engagement of the faculty member, student, and disability services staff in the discussion.

Simulation learning as a course/program component

The second application of simulation learning for students with disabilities occurs when the simulation exercise is a core part of the course or program the student is participating in. In this scenario, the accessibility of the simulation is of paramount consideration. It is necessary to check with the student using the simulation learning exercise about whether it is fully accessible, and consider whether other, more accessible, simulations would be more appropriate, if they meet the learning objectives to the same degree.

Many of the concerns around accessibility of simulation learning are similar to those surrounding online discussion forums and learning spaces (e.g., Blackboard and Moodle). It is worthwhile, in these situations, for the faculty member to consult with the disability services staff as well as staff from campus teaching and learning centers who may be experienced with new learning technologies and international software accessibility compliance guidelines.

Simulation learning in postsecondary education

Many resources and repositories for simulation learning in the sciences exist online (e.g., PhET Interactive Simulations from the University of Colorado: https://phet. colorado.edu/). Some simulations are designed very specifically for the elementary and secondary school settings, while others delve deeper into concepts from a variety of fields at the university and college levels.

It is worth noting that given the nature of graduate education and training in the sciences, it is not feasible to utilize simulation (specifically virtual learning) approaches at the graduate level outside of the classroom setting. Fieldwork and experimental labwork cannot be easily or feasibly supplanted by or replaced by simulations. However, an exception to this rule exists in fields where the simulation process itself can be incorporated into the student's thesis project (e.g., mathematical modeling, artificial neural networks, etc.), thus contributing in a potentially meaningful way to the acquisition of knowledge.

Conclusion

Simulation learning is increasingly becoming a valuable (and creative) teaching tool at the secondary and postsecondary education levels. At the elementary level, simulation learning might be a valuable and accessible way to expose young students (with and without disabilities) to STEM—e.g., NASA's Q&A sessions and experiments from the International Space Station. Simulation learning as an accommodation approach needs to be evaluated in terms of accessibility of the simulation as well as degree of concordance between the learning objectives of the simulation and the essential requirements of the course or program. Meanwhile, simulation learning as a mainstream teaching tool needs to be evaluated for its accessibility and adherence to universal design principles. Alternatives should be considered if these criteria cannot be met or determined. Simulation learning lacks a direct "hands-on" component that other students will obtain through participation in the lab or exercise, and deployment of this tool as an accommodation needs to take into consideration the student's preferred learning style. As with technical assistance and other forms of accommodation, simulation learning should be evaluated for its fit with the student, and the student, faculty, and disability service provider all need to be full participants in the discussion.

Physical access in science laboratories

Chapter Outline

Introduction   258
Considerations for physical accessibility in science labs   259
Universal design and physical accessibility of science laboratories   259
Best practices   260
Accessibility and safety   261
Conclusion   262

Science labs are busy places.

I don't mean work-wise – although there's that, as well. A lab with every workstation occupied is a sight to behold, a bustling beehive of activity, in many ways frenetic, controlled chaos. Also a nightmare to navigate, but we'll come back to that....

No, I mean that science labs are often full of stuff. Equipment on benches. Power cables running every which way. Bottles of hazardous chemicals and hazardous chemical waste side by side in fume hoods. Supplies under workstations, on shelves, in cabinets, crammed, essentially, wherever they will fit. Solutions, chemicals, plasticware on racks, on workstations, on shelves as well as various handwritten notes and memos found posted in odd places everywhere. The science lab is built for people to stand in – or, sit in very high chairs. A lot of labs are, in fact, built for basketball players – step ladders, the sort that you find in a kitchen or larder, are surprisingly common.

Science labs are busy. Space is often at a premium – indeed, the final frontier. When it comes to placement of equipment or supplies, in many working science labs, anything goes, as long as it doesn't compromise the safety or emergency response regulations, and even then sometimes productivity comes before conventional wisdom.

All this comes to be because there's only so much square footage, and often universities, colleges, and research institutes alike all want to maximize the number of working scientists in the space. Couple that with a simple fact that I've discovered in my scientific career: Scientists are packrats. We almost never throw anything out. So, labs are small and accumulate clutter over time.

Newer research spaces look bigger and airy, but many of those are built on the open concept model, where everyone shares everything.

Creating a Culture of Accessibility in the Sciences. DOI: http://dx.doi.org/10.1016/B978-0-12-804037-9.00023-1

If a fully staffed and functional lab with all workstations occupied is a beehive of activity, imagine ten of them all crammed into the space that should be allotted for five. And there being no doors, walls, partitions, or other dividers.

It's an introvert's worst nightmare – and a navigation hazard for someone with a visual impairment. Aisles are narrow, so trying to get around people and equipment without bumping into anything is next to impossible. Trying to find what you're looking for in a space where it's hard to impose a common organizational scheme, and labels which are difficult to read, takes a long time.

Lighting is stark. Glare and contrast can be a problem. Most labs I've been in are too dim because the overheads are poor, or are shaded by boxes on the highest shelves—or, they're too bright. Perversely, the shelving can cast shadows onto the workstations in visually unfriendly ways, making it hard to pick materials out of those areas.

In all this, a working scientist is allotted a miniscule three feet (one meter), at most, of bench space in which to work. Depending on how crowded the environment is, they may have to share that with a colleague.

Never mind the inaccessible sinks or fume hoods or workstations – also big issues – a traditional science lab is overwhelming and visually inaccessible on many fronts.

If you're an experimentalist, this is what you're working in – it almost doesn't matter the field.

Wouldn't it be nice if someone tried to design, without any assumptions, what a truly accessible science lab would be like?

Introduction

Discussion of the physical accessibility of science laboratories remains one of the most difficult challenges faced by students with disabilities in science, technology, engineering, and mathematics (STEM) fields, their faculty, and disability services staff. From teaching labs in colleges, undergraduate settings, and professional programs, to working research labs encountered by graduate students and postdoctoral scholars, the range of science learning, teaching, and research environments presents significant barriers for trainees interested in pursing STEM training. While we tend to focus on the physical structure of the lab for someone in a wheelchair, it is worth noting that the lab environment presents many visual accessibility challenges as well as difficulties for individuals with chronic disabilities.

At best, the physical structure of most laboratories is unwelcoming to people with physical disabilities; at worst, it is inaccessible. Many laboratories are difficult to navigate and visually obstructive. Lab spaces are often encumbered by high workbenches, inaccessible cabinets and shelving, and overcrowded fragile equipment (Hilliard et al., 2011). Faucets for sinks, gas hookups, power outlets, fume hoods and biological safety cabinets, eyewash stations, and other safety equipment also pose difficulties, because these areas of the lab are not easily accessible to persons unable to stand.

Although there are some significant best practices, which we will highlight here—including a trailblazing checklist on developing accessible science lab environments (Doyle, 2014)—the STEM literature has not developed a standard design for an accessible science lab in any field (Moon et al., 2012). The literature also provides little in the way of examples for accessible laboratory equipment used in the instruction of STEM courses at the postsecondary level (Moon et al., 2012). In this chapter, we present an overview of some of the key considerations that faculty, disability services staff, and students need to think through when planning the accessibility of a science lab, and highlight some best practices that illustrate physically accessible science lab environments.

Considerations for physical accessibility in science labs

The recently completed Science Building at the University of Toronto at Scarborough (Toronto, Canada) holds the distinction of being the first building with science labs constructed according to universal design and accessibility principles and based on a checklist developed by Doyle (2014). The lab spaces were designed and constructed to be in compliance with a comprehensive physical accessibility checklist developed internally (Doyle, 2014), and the checklist included a number of considerations of relevance to persons with disabilities needing to access a science lab, and was developed from universal design principles for physical spaces.

Considerations for the checklist included: Space and reach requirements; accessible routes within the building and paths of travels, accessibility of doors between rooms, ground and floor surfaces, controls and operating mechanisms, demonstration areas, workstations and furniture, fume hoods, sinks, emergency showers and eyewash stations, lighting, and signage. The checklist was developed with the potential barriers faced by persons using mobility aids such as wheelchairs or scooters in mind, along with the potential barriers faced by people with disabilities for whom navigation, reach, and standing for long periods may be a challenge (e.g., persons with visual disabilities or persons with chronic disabilities).

Universal design and physical accessibility of science laboratories

As noted earlier in this book, universal design is "the design of products and environments to be usable by all people, to the greatest extent possible, without the need for adaptation or specialized design" (Center for Universal Design, 1997).

First implemented in the design of buildings, public spaces, and products, universal design has been hailed as cost-effective, because it seeks one integrated approach to accommodate diverse user characteristics, and addresses the needs of many people without stigmatizing any group.

The physical accessibility of a lab impacts whether students with disabilities are able to succeed. Physical accessibility can pertain to all aspects of the lab environment, specifically in relation to workbenches, sinks, fume hoods, showers, and eyewash stations. The fact that there are best practices in making labs accessible is, in itself, evidence that supports the requirement to critically assess the physical accessibility of standard laboratory design. Supporting the independence and inclusion of persons with disabilities in the laboratory requires providing a safe, accessible environment with practical assistive solutions (Hilliard et al., 2011). A review of the literature reveals examples of laboratories in Canada (such as the iScience lab at McMaster University) and laboratories in the United States (such as the Accessible Biomedical Immersion Lab, or ABIL, at Purdue University) that have been constructed to accommodate students with physical disabilities (Hilliard et. al., 2011).

Best practices

One initiative designed to increase opportunities for students with disabilities is the construction of the Accessible Biomedical Immersion Laboratory (ABIL) at Purdue University located in Lafayette, Indiana. The purpose of this project is to enable persons with physical disabilities to access standard laboratory and safety equipment. Modifications to a science laboratory at Purdue University's Discovery Learning Research Center were made to facilitate scientific research for students with disabilities in a learning lab environment (Hilliard et al., 2011).

A similar initiative, the iScience (Integrated Science) laboratory, has been undertaken at McMaster University located in Hamilton, Ontario, in the General Sciences program. This accessible lab opened in September 2013. The iScience laboratory is designed to teach numerous disciplines, including life science, physics, and chemistry. All utilities are in place in an integrated setting, with safety features fully considered. An excellent YouTube video on the McMaster program is available.

The construction of ABIL was centered on the architectural features required to perform wet bench research. Several modifications to the lab at Purdue University were made to enable students with physical impairments to easily access equipment. The ABIL project itself was initiated by Prof. Brad Duerstock, a scientist with a spinal cord injury who specializes in the fields of neuroscience and assistive technology design. The lab design was based on Duerstock's personal experience and expertise in the field. Duerstock also founded and leads the Institute for Accessible Science. In the Purdue laboratory, the lab bench, fume hood, and sink were identified as three main tools that students would need to access to perform hands-on scientific research. These tools were placed close to each other to enhance a student's efficiency in the lab, but also ensure a student has the necessary room to maneuver (Hilliard et al.,

2011). A traditional lab bench was replaced with an adjustable version that accommodates researchers with mobility disabilities. To improve access to needed equipment, several commonly used pieces of equipment were placed in close proximity to the adjustable lab bench, including a talking scale for low-vision users, motion-activated biohazards waste containers for researchers with mobility impairments, and an automated paper towel dispenser (Hilliard et al., 2011).

To ensure the accessibility and usability of the sink, the counter height was lowered to accommodate students who are seated in wheelchairs or who are short in stature (Hilliard et al., 2011). The faucet neck and handles were placed parallel to one another near the front of the sink, and the area below it was cleared to accommodate wheelchair users (Hilliard et al., 2011).

This example illustrates that many factors must be taken into account when considering the physical accessibility of a space. To make a space accessible to everyone, it is essential to consider table heights, open spaces for maneuverability, sink heights, table heights, object placement, and other issues. Other elements that can help to increase physical accessibility include mirrors above demonstrations, enlarged screens, lowered controls, electric stirrers, extended eyepieces for microscopes, and modified procedures. As with any accommodations pertaining to disabilities it is best to consult the science student with a disability or the professional scientist with a disability to find out what they need for success in the lab.

Accessibility and safety

Safety is perhaps the biggest concern raised by persons with disabilities in the lab environment—safety in reach and handling of equipment; safety in handling dangerous physical, chemical, or biological substances; safety in the event of an emergency. Indeed, it has been reported that the perception of safety risks "could inhibit enrollment by students with disabilities in chemistry laboratory courses" (McDaniel et al. 1994, p. 21). However, it is worth noting that often these decisions are made by faculty, in the absence of consultation with the student, during admissions review processes, thus fitting our definition of the "gatekeeper function" (see chapter 3: Barriers faced by students with disabilities in science laboratory and practical space settings; and chapter 10: Essential requirements and academic accommodations in the sciences). It is also worth acknowledging that an accessible lab, designed according to universal design principles, is also a safe lab (Blake-Drucker, 2009).

Students with disabilities are often very familiar with their needs, strengths, and limitations. As such, Jones (2002) argued that these students are "more safe ... and are inclined to be more deliberate and forward thinking when doing particularly unusual or extreme experiments." Indeed, the prepared student has often evolved strategies to manage their safety and the safety of those around them in the lab environment—the more experienced students are, and the more advanced their training, the better prepared they are. This perspective argues for the active inclusion of the student in honest and transparent discussions around lab safety. Indeed, it is imperative that safety be discussed with each student with a disability, "including instructions for preventing and responding to situations that would compromise the safety of the student or others" (Doyle, 2014). A separate "accessible safety tour" of the lab for the student may be warranted, highlighting exit routes, how to access spill kits, location and operation of eyewash stations, procedures for handling dangerous equipment and/or chemicals, and what to do in the event of a chemical spill, chemical exposure, or injury.

Safety in the context of a mobility aid (e.g., what to do if a chemical spills on a wheelchair), service animal (e.g., how to ensure the service animal's safety in the lab, and what training may be required for both lab members and service animal in the context of safety), or human assistant (e.g., safety training for a sign language interpreter) also needs to be considered (Doyle, 2014). It is also crucial to ensure staff and faculty responsible for safety in laboratories are familiar with the needs of students with disabilities, and that they are aware of the institution's key contacts (e.g., in the disabilities services office and in health and safety/risk management) for consultation when required (Doyle, 2014).

A crucial subset of the safety conversation is related to emergency preparedness (e.g., evacuation planning and fire response). It is important to review and discuss the accessibility requirements of the student in the event of an emergency and to identify and designate appropriate assistance from the lab staff as necessary. Once discussions around accessibility and emergency preparedness, which include the student, have been held, a written emergency preparedness plan should be developed and shared with the student, faculty supervisor of the lab, and appropriate lab staff, and required training of lab staff be undertaken to complement the plan.

Conclusion

Of course, planning for an accessible science lab can be expensive. However, retrofitting a science lab to make it accessible also has significant costs—financial, productivity, and time. Furthermore, an accessible science lab also takes all staff members into account. Various features of an accessible science lab are also beneficial for the 10% of the population that is left handed, pregnant women, staff members carrying heavy material, as well as individuals of differing heights and strengths. When it comes to physical accessibility of a science lab, safety, accessibility, and universality are truly synonymous concepts.

Part VII

Synthesis

Practicum placements

Chapter Outline

Introduction: overview of practicums   266
Review of relevant legislation and duties to accommodation   267
Introduction to the practicum accommodations process model   268
Contextual considerations   270
    National context   270
    University context   270
    Practicum site context   271
    Student context   272
Partnerships required for successful practicum placement learning for
    students with accommodations   272
    Student   273
    University faculty   273
    University services   273
    Practicum site   274
Practicum accommodations process   274
    Initial meeting of student and course instructor to review expectations and explore options   274
    Accommodations assessment/recommendations   276
    Explore potential practicum partners   276
    Finalize specific practicum arrangements   277
    Monitor performance throughout practicum   277
    Reflect on teaching and learning   278
Considerations and strategies for successful provision of practicum
    accommodations   278
    Maintenance of academic and professional standards   279
    Time   280
    Equitable opportunities for all students   281
Student accommodations   281
    Altered time   281
    Geographical location   281
    Provision of one primary preceptor   282
    Homogenous practice area   282
    Use of technology   282
Considerations for national and international practicums   282
Conclusion   283

*Chelsea's Story: Fieldwork and practicum placements are challenging
waters for any student to navigate; you find yourself in an unfamiliar
environment, having to apply the practical components of what you
learned in lecture. However, these challenges are, in my experience,
heightened for students with disabilities. I say this not because it is*

Creating a Culture of Accessibility in the Sciences. DOI: http://dx.doi.org/10.1016/B978-0-12-804037-9.00024-3
© 2017 Elsevier Inc. All rights reserved.

any more difficult for us to transfer the in-the-classroom knowledge to placement, but because of the innumerable attitudinal barriers we face in trying to enter the fields of our choosing.

I think back to the time when I had just completed my undergraduate degree and was considering what field I'd like to pursue for graduate school. I had strongly considered social work, given my interest in working with, and mentoring, youth with disabilities. Upon acceptance to the social work program, I met with the fieldwork/practicum advisor to discuss the accommodations I would require for my placement. I explained my career goals, and provided some potential suggestions for placements.

I recall being informed at the outset that "an inability to drive would make a career in social work virtually impossible." At this point in the conversation, I explained that I had researched some potential placements and career options that likely would not require me to drive. One such option I presented for a potential placement was working in a hospital setting. The response I got was "well, hospitals are very big, intimidating places, even for those of us who can see."

At this point in my educational journey, I admit I was not fully aware of all of my options had I opted to remain in the social work program and pursue the placement. What I did know, however, is that the work I wanted to do was important to me, and there was more than one way to do it. At this point in my academic journey, I had to decide whether to remain in the social work program and break down the barriers to my practical learning, or enroll in an alternate program where the barriers would no doubt still exist, but be more manageable. I opted to self-select into a research-based master's program, which provided an excellent platform to address disability-related issues within a profession-specific environment. In addition, I used this learning opportunity to build valuable contacts and networks within the disability community, which resulted in opportunities to engage in leadership and mentorship roles.

Unfortunately, I did not push the field coordinator to think more critically about being creative in an approach that would work for me to do the kind of work I wanted.

Instead, I opted for an alternative route.

Introduction: overview of practicums

Practicum placements are out-of-classroom learning opportunities that are integral components of many curricula. Numerous professional programs such as healthcare and education require practicum placements to ensure student application of class-room knowledge, integration of theory with practice, and the development of pro-fessional competencies. Professional associations and/or accreditation bodies often prescribe the required number of hours for student practicums to ensure adequate opportunity for the acquisition of required competencies to practice. In addition to mandatory professional placements, other areas of academic study provide practicums to enhance student learning.

Practicum placements must be accessible to all learners, including those with disabilities, to ensure all students have opportunities to realize the many benefits of these rich learning experiences. Practicum placements provide opportunities for students to network with and learn directly from professional role models and to develop their own professional competencies to practice as they are being "socialized" into the profession (Dornan & Bundy, 2004). In addition to discipline-specific knowledge, skills, and behaviors, students learn general work skills such as working collaboratively with others, communicating effectively, prioritizing daily and weekly tasks, and taking responsibility (Hall, Healey, & Harrison, 2002). Practicum placements can aid in students' employment search after university through professional and personal development and networking. In addition, students' exposure to different work settings and responsibilities provides an opportunity for them to further understand and match their own goals, interests, and abilities to work environments/organizations (Sharby & Roush, 2008). For students with disabilities, practicum placements provide an excellent opportunity to address disability-related issues within a profession-specific practice setting and to advocate for their needs (Johnson, 2000).

Practicum placements, also known as "fieldwork placements," "clinical education," "field-based learning," "internships," "experiential education," or "service learning" may differ in setting, goals, learning objectives, and duration. Although practicums with an additional layer of complexity, such as those that take place outdoors and involve overnight stays and meals, are not specifically addressed within this chapter, the general principles and processes that are discussed should be helpful for administrators, preceptors, and instructors of these courses and programs. Research experiences in laboratories and in the field, by contrast to practicum settings, have been discussed in previous chapters; this chapter focuses specifically on the practicum setting. Throughout this chapter, the facility where the practicum placement is taking place will be referred to as the practicum site, the practicum instructor will be referred to as the preceptor, and the university faculty member teaching the course will be referred to as the course instructor.

This chapter is written with the social model of disability lens (Gill, 1994). Students may be prevented from equitable participation in practicum experiences due to social, environmental, and attitudinal barriers. While the student must advocate for themselves, the focus should not be on their physical and psychosocial impairments.

Review of relevant legislation and duties to accommodation

Acts and laws have been passed in many countries to ensure individuals with disabilities have their rights protected and are free of discrimination. These national acts are country-specific reflections of the United Nations Convention on the Rights of Persons with Disabilities (see chapter 2: Accessibility and science, technology, engineering, and mathematics—the global perspective). For example, in Canada, this legislation includes human-rights laws across all provinces such as the Ontario Human Rights Code (Ontario Human Rights Commission, n.d.), the *Accessibility for Ontarians with Disabilities Act* (AODA) (Ontario Ministry of Community and Social Services, 2005), and federally, the *Canadian Charter of Rights and Freedoms* (Department of Justice, 1982). In the United States, federal legislation similarly protects individuals with disabilities from discrimination through the *Americans with Disabilities Act* of 1990 and Section 504 of the *Rehabilitation Act* through the modified *Americans with Disabilities Amendments Act* (ADAA, 2008). In the United Kingdom, individuals are protected from discrimination through the *Equality Act 2010* (n.d.) as well as the *Disability Discrimination Act* (1995) and the *Special Education Needs and Disability Act* (2001).

With regard to human rights as a whole, the United Nations Convention on the Rights of Persons with Disabilities helps not only to protect individuals but also to promote their rights. University personnel including those at the disability services office, course instructors, and preceptors have a duty to abide by applicable laws and acts as well as university policies. How these acts, laws, and policies apply to students can be difficult to understand and yet they need to be carefully understood and followed so that students can equitably participate in practicum placements.

Introduction to the practicum accommodations process model

Individualized accommodations may be necessary for students to equitably participate in practicum placements. The Practicum Accommodations Process Model (PAPM; Diagram), describes the factors and processes involved in making decisions and arrangements for student accommodations in practicum placements. This model builds on the Sharby and Roush (2009) "six-step collaborative decision-making model" that was developed to assist with the implementation of student accommodations in the American allied health education arena. Components of that model include:

1. Positive climate
2. Essential functions
3. Student challenges and strengths
4. Learning activities
5. Reasonable accommodation
6. Ongoing assessment

The PAPM differs from the Sharby and Roush (2009) model in that it specifically outlines the accommodations process for practicum placements, it further expands on the complex contexts in which the accommodations process takes place, it illustrates the complexity of the essential communication that takes place between partners, and it outlines the specific steps of the accommodations process. The PAPM is applicable for students with any disability, in any context, and in any discipline that requires practicum placements.

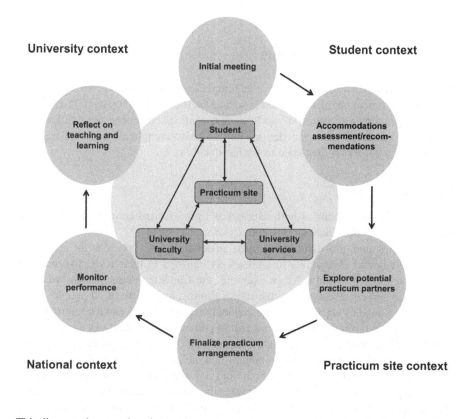

This diagram shows various interactions of student, practicum site, university faculty, and university services. A circular flowchart surrounds the interaction. The interactions are shown with double arrows and are between the following: Student and Practicum site; Student and University faculty; Student and University services; University faculty and University services; and Practicum site and University faculty. The circular flowchart begins with "Initial meeting" leading to "Accommodations assessment or recommendations," to "Explore potential practicum partners," to "Finalize practicum arrangements," to "Monitor performance," which finally leads to "Reflect on teaching and learning." The area between "Initial meeting" and "Accommodations assessment or recommendations" is labeled "Student context." The area between "Explore potential practicum partners" and "Finalize practicum arrangements" is labeled "Practicum site context." The area between "Finalize practicum arrangements" and "Monitor performance" is labeled "National context." And, the area between "Initial meeting" and "Reflect on teaching and learning" is labeled "University context."

The outer area represents the various contextual factors that have an effect on the student accommodations process and that aid in decision-making about practicum placements. The inner circle illustrates the complexity of communication that must occur between the four partners throughout the accommodations process.

The multiple steps of the student accommodations process that involve the four practicum partners and take into account the contextual factors are shown in the outer circles. The model is explained in more detail later.

Contextual considerations

The outer area of the PAPM lists four contextual factors that must be considered when preparing for practicum placements. These factors, the "national context," "university context," "practicum site context," and "student context" can encourage and facilitate the process of student accommodations but may also complicate and challenge the process. Each context is explained in further detail later.

National context

The national context includes both national and professional components that guide and direct student accommodations in practicum placement.

Universities and practicum sites must understand and consider their national cultural and lawful obligations to students as outlined in their respective national disability-related legislation. As an example, the Canadian culture is known to be inclusive and diverse, recognizing the rights, dignity, and worth of all people within the country. Canada's values and beliefs related to disability and equal opportunity are formally voiced through legislation regarding persons with disabilities as previously discussed. Canadian law requires that students with documented disabilities have the right to accessible learning opportunities, which includes practicum placements.

In addition, laws that pertain to both students and recipients of student services within the practicum site must be considered in the placement accommodations process. For example, privacy and confidentiality laws must be observed when in communication with the practicum site about the need for student accommodations and also when considering appropriate accommodations within certain environments where the student is learning and providing services. For example, accommodations including video or voice recording of patient/client sessions may not be possible in all settings due to privacy legislation.

If practicum placements are a required part of a professional academic program, professional practice standards, codes of ethics, and requirements from regulatory colleges and professional organizations and associations must also be taken into consideration when assigning alternate learning experiences or student-specific accommodations for practicum-based learning. Professional standards and requirements cannot be compromised during the accommodations process in terms of educational/training or practice standards for the profession or client care and services (Pardo and Tomlinson, 1999).

University context

Universities and colleges strive to provide inclusive, accessible, and equitable learning environments. Practicum placements, as part of an academic program or coursework, must conform to published policies, including academic standards and associated relevant legislation. The university context takes into account the policies and procedures pertaining to course development, structure, and delivery of academic performance and student accommodations that must be followed. The university context also includes faculty and staff who have a role in the provision of accessible education. The student accommodations process for practicum placements is assisted by a variety of university support services that are available for students such as individualized assessment, counseling, and health services. Students with disabilities can and should draw upon these resources to facilitate appropriate accommodations and support throughout the process.

University and college course calendars should inform students of practicum placements and/or requirements within a course or program of study. Course outlines should give further information such as any additional costs or flexible practicum hours and requirements such as immunizations and police checks. The outline should include a statement regarding reasonable student accommodations and the guidelines and processes to speak with the course instructor and/or university student services about potential accommodation needs (Dupler, Allen, Maheady, Fleming, & Allen, 2012).

The course instructor should carefully recruit appropriate practicum sites for the practicum learning experience. In addition to adhering to university requirements such as legal agreements with practicum sites and student insurance coverage for accident or injury, course instructors should remind sites that practicum placements need to be accessible for all students. It has been suggested that the contract between the university and the practicum site include a statement that speaks to compliance with antidiscrimination laws (Kornblau, 1995). Course instructors should aim to recruit a variety of settings to maximize student interest, accessibility, and participation.

Preceptors within the practicum sites must be prepared for their teaching role to ensure student learning goals are accomplished and academic standards are upheld. Preceptor education should include clarity around student learning objectives, how to minimize not only physical but social and attitudinal barriers, and the methods and tools for evaluation. Instruction on teaching techniques such as the provision of learning opportunities appropriate to the students' education level and giving helpful feedback should also be offered as required (LoSciuto, Rajala, Townsend & Taylor, 1996).

Practicum site context

Learning opportunities and practice requirements will vary among practicum sites. Practicum placements occur within organizations (e.g., hospital, school, industry, private practice) that must follow relevant national legislation and that have developed unique policies and procedures. Students learning within the organization must comply with all applicable organizational policies, procedures, and requirements such as police checks, health and safety requirements, and site-specific training so they will not compromise the organization's mission or reputation nor endanger clientele, themselves, or other staff. Thus the practicum site context influences the need for and provision of reasonable accommodations as the impact of a student's disability can shift depending on practicum site requirements (Cooley & Salvaggio, 2002).

Practicum sites will have varying abilities to provide necessary reasonable student accommodations based on the nature of the organization's work, their clientele, and the accessibility of the physical facilities. The physical, psychosocial, and cognitive demands on the student should be understood for each practicum site and/or practice context. To assist with the practicum site assignment of a student with accommodations, a formal analysis that addresses the physical, programmatic, informational, and attitudinal accessibility of the practicum site and its ability to accommodate experiential learners with disabilities could be implemented as required (Chelberg, Harbour, & Juarez, 1998; Johnson, 2000). Student accommodations could then be more easily matched to a suitable practicum site. The practicum site demands and preceptor expectations of the student should be thoroughly discussed with the student to ensure adequate preparation and resources (Galvaan, 2012).

Student context

Student context encompasses the strengths and challenges of the student and includes the physical, cognitive, and psychosocial characteristics of the student in addition to available resources such as friends, family, other professionals, equipment, finances, etc. The course instructor and other university faculty may or may not initially know of these personal strengths and challenges but through ongoing conversations with the student will become more aware of the student context. The student has the greatest understanding of his or her capabilities and will likely have had experience involving necessary accommodations in many different situations throughout his or her life (Beckel, 2012).

Resource strengths include personal support systems and material resources. Students with disabilities may have a community of personal supports such as friends and family in addition to already established nonuniversity professional supports such as counseling and health services. A student support community is a great resource for the student during all aspects of the accommodations process. This community often provides physical, financial, and emotional support and encouragement for the student. Although a vital aspect of student success, members of the student's personal support community do not usually become formal partners in the practicum accommodation process. Resource challenges refer to the lack of personal supports in the way of people, services, and finances a student has access to. Finances can be an important consideration for students as some may require medications and equipment to ensure optimal day-to-day functioning. Financial strain can increase stress and affect student performance.

Partnerships required for successful practicum placement learning for students with accommodations

The middle circle of the PAPM identifies the key practicum partners and shows the lines of communication that occur during the student accommodations process.

Interdependency and clear communication between partners is essential during the entire practicum placement process. The student and university faculty communicate with university services and the practicum site, but the practicum site only communicates with the student and university faculty. Communication among the partners is ongoing throughout the process and is vital for successful accommodations of students in practicum placements. Partnerships are further examined later.

Student

The student is the primary partner in the accommodations process. Students requiring accommodations are usually keenly aware of their personal strengths and challenges and are often able to provide creative solutions to practicum placement issues. Students should be encouraged to take a leadership role in the accommodations process, although they may benefit from ongoing coaching from the course instructor in order to advocate for themselves and approach discussions with the preceptor regarding challenges that arise during the learning experience. Students must be effective and yet respectful advocates for their own learning as they communicate with all practicum partners.

University faculty

University (and college) faculty includes those who are within the student's program of study and are actively engaging in discussions with the student regarding required accommodations for practicum placements. These include the course instructor and other related faculty such as a vice dean or registrar, graduate student coordinator, or faculty advisor. The course instructor will communicate with the student concerning day-to-day coursework, while faculty overseeing placement and practicum may have a more long-standing working relationship with the student, to ensure that equitable and accessible learning occurs throughout the student's program of study.

It is highly recommended that a main university contact person be assigned to assist students in the referral to student services as well as liaise with the practicum site (Dupler et al., 2012; Howlin, Halligan, & O'Toole, 2014). The course instructor is most often the assigned contact person. Clear and frequent verbal and written communication is required before, during, and after the practicum. The course instructor is the only other person, besides the student, that may be communicating with all other fieldwork partners. The role of the practicum course instructor has been described as a "balancing act" as this person not only is considered an agent of the academic institution and an advocate for the student but must also consider the needs of the practicum site and the preceptor (Beckel, 2012).

University services

University (and college) services include resources and services that can be accessed by the student throughout his or her time of enrollment. These can include health and wellness services that offer medical and diagnostic services, personal counseling, peer counseling, and health-promotion services. The student may also utilize academic and writing support services. The disability services office provides individual assessment of disability-related accommodations and resources for the student and communicates with faculty within the student's program. It is helpful if the course instructor, student, and the student's accessibility counselor communicate several weeks before the start of course to discuss reasonable practicum placement accommodations and how they will be implemented.

Practicum site

Partners at the practicum site most often include a student coordinator and the student's preceptor. Other staff such as an occupational health and safety officer may also become involved as required. To work efficiently and ensure clear communication, it is recommended that all communication from the university come from the assigned main contact. It is also recommended that whenever possible, communication with staff at the practicum site is documented by the university and shared with the site. The practicum site will also be in regular communication with the student once the student has been assigned to the facility.

Practicum accommodations process

A structured process can help to ensure students requiring accommodations are provided with equitable and accessible learning opportunities. The process for the provision of accommodations can be quite complex due to the number of people involved and the varied settings. This highlights the need for careful planning, communication, monitoring, and reflection during and throughout the practicum placement. The student should be encouraged to contact the assigned university faculty member well in advance of the practicum placement opportunity to ensure all steps of the process are followed: expectations are reviewed, options are explored, referral for student accommodation assessment and recommendations are made, practicum partners are explored, and arrangements are finalized. Student performance throughout the experience should be monitored to ensure his or her learning needs are met.

Initial meeting of student and course instructor to review expectations and explore options

As suggested earlier, it is recommended that the student contact the course instructor well in advance of the practicum placement start date in order to discuss the learning opportunity. By advocating for themselves in a timely manner, the student can help to ensure his or her learning needs are met. The student knows what he or she is capable of and can provide valuable insight into the specific support required (Howlin et al., 2014). Disclosure of a disability is not required but disclosure of specific learning needs for the practicum placement is considered as a professional student responsibility.

With early contact, the course instructor can work with the student to provide necessary information about the practicum placement learning opportunity and work with the student to identify any potential challenges that may arise during placement and problem-solve solutions ahead of time (Kornblau, 1995). During the initial meeting, it is important for the course instructor to create a safe and welcoming environment so that feelings of isolation, fear, or separateness are minimized (Dunn, Hanes, Hardie, Leslie, & MacDonald, 2008; Graham-Smith & Lafayette, 2004). To help reduce student anxiety, it is also helpful to provide detailed information on what the student might expect on the practicum, e.g., what a typical day might look like and who the student would be interacting with (Healey, Jenkins, Leach, & Roberts, 2001). The student should also be assured that the course instructor can be readily contacted if issues arise (Dupler et al., 2012).

It is also important to discuss the student's learning needs to ensure that the course instructor does not make assumptions about what those needs are. During the initial contact, the course instructor and student should also discuss relevant consent and confidentiality issues. In addition, discussions could also include appropriate university resources and personal supports such as physician, disability services office, health services, faculty advisor, etc.

The course instructor should review the practicum learning objectives and performance expectations with the student, keeping in mind that the academic standards for the program cannot be compromised. The practicum schedule including required daily attendance (full-/part-time) as well as start and end times needs to be discussed. Additionally, initial discussions should include a review of the required competencies and skills that must be met within the stated timeframe of the practicum. Within this discussion, the expected level of student independence should be highlighted. Fortunately, many practicum sites provide accommodations to ensure students can equitably participate and gain the required competencies and skills required by their program.

In some cases, students are allowed to submit practicum site/learning opportunity preferences that are taken into consideration during the student practicum placement assignment process. The student should be encouraged to openly discuss these preferences with the course instructor in relation to his or her individual learning needs and the need for accommodations. It is often best to receive the accommodations prior to the assignment of the practicum placement.

The student may wish to discuss the question of whether, when, and how to discuss accommodations with a preceptor due to issues such as stigma, negative attitudes, and being singled out among students (Francis, Salzman, Polomsky, & Huffman, 2007; Howlin et al., 2014). The fear of discrimination is real and can lead students to feel labeled and disempowered (Wray, Fell, Stanley, Manthorpe, & Coyne, 2005). A course instructor should respect the student's choice if he or she elects not to pursue any accommodations for his or her practicum placement, but should advise students of any potential foreseen challenges resulting from not requesting needed accommodations. However, considerations of the health and safety of the student and individuals they will be working with need to be addressed and can be best done if the student discloses the need for practicum accommodations. There is the potential for lawsuits (Reeser, 1992) so this conversation in the initial meeting(s) is encouraged.

Accommodations assessment/recommendations

The course instructor should refer the student to the university's disability services office in a timely manner so the student can be assessed and the provision of practicum-specific accommodations can be identified. Discussions between the course instructor and student and between the course instructor, student, and disability services office personnel should be ongoing, keeping student privacy in mind. The course instructor may need to educate the accessibility counselor regarding specific details of the practicum experience such as the physical, cognitive, and psychosocial demands of the student to ensure that accommodations are reasonable and appropriate for the experience. A clearer understanding of the student's abilities and required accommodations can be informed by the student's active participation in simulations and labs prior to the practicum (Azzopardi et al., 2014; Francis, et al., 2007). Once accommodations have been identified, the course instructor should be provided with the recommended accommodations well ahead of the practicum start date to ensure accommodations can be implemented in a timely manner.

Explore potential practicum partners

The learning needs and accommodations of the student are considered in relation to available practicum placement sites. The practicum site's essential functions should be compared with the student's required accommodations to determine the feasibility of student success (Francis et al., 2007). Additionally it is important to remember that students with disabilities may not require accommodations within all learning environments (Sharby & Roush, 2009), so a clear understanding of the site's essential functions, expectations, and environment is essential.

It is often beneficial to explore appropriate placement options for students with disabilities prior to the practicum site assignment being made available to all students to ensure the maximum number of practicum placement options are available for students who may require accommodations on-site. The course instructor should contact appropriate practicum sites to discuss the student's accommodation needs and ensure preceptors know how to effectively accommodate for the student's learning needs. In our experience, students have reported that it has been helpful when potential suitable practicum sites were identified in advance and accommodations were discussed and implemented rather than placing them at any site and then trying to make it work after the fact. In this way, the site knows what to expect, and can plan accordingly, which in turn helps the student feel more valued and confident in his or her abilities. Some preceptors may have concerns about the additional time and effort it may take to plan, arrange, and monitor student accommodation (Sowers & Smith 2004) and these concerns need to be addressed. Other preceptors may note the existence of resource barriers including staff shortages or environmental barriers that would need to be addressed in order to effectively implement specific student accommodations.

When identifying potential practicum sites, students and preceptors have emphasized that the attitudinal barriers can play a significant role as to whether or not the site is accommodating (Sowers & Smith 2004). These sentiments were similar to those of Beckel (2012), where students with disabilities in professional programs expressed that the greatest barrier toward success in an academic program came from the attitudes of others around them. In our experiences, students have emphasized how difficult it was when university personnel and preceptors did not believe they needed accommodations. Some students noted that they felt accommodations were sometimes viewed as trying to get out of doing work. Preceptors who are inclusive, receptive, flexible, creative, and willing to problem-solve with students can truly optimize the learning experience.

In an ideal world, all practicum sites would be accessible to all students, but in reality, many still are not. The exploration of practicum partners is a critical step to ensure equitable learning.

Finalize specific practicum arrangements

An appropriate practicum placement is then assigned to the student that takes into account the student's required accommodations. If there are associated costs for the provision of any equipment or software, funding must be secured with appropriate deadlines for purchase and implementation prior to commencement of the practicum. As details are finalized, the practicum site should review relevant legislation and site policies and procedures to ensure students can safely and equitably participate in the placement. The student should be actively involved in discussions with the course instructor and where appropriate with the practicum site around the practical aspects of the practicum placement and how the accommodations will be implemented. Including students in conversations with practicum sites also provides them with valuable experience for requesting accommodations in the future and advocating for themselves (Beckel, 2012; Johnson, 2000). Creative solutions are found for many obstacles when practicum partners are flexible and have a positive attitude toward equitable student learning opportunities. Involving the student and practicum site to creatively develop strategies can prove to be satisfying for all involved (Dupler et al., 2012). In some cases, it is advisable for the student to contact the site prior to commencement of the learning opportunity so that an early introduction can help to ensure required site requirements and responsibilities are discussed (Sharby & Roush, 2009) and an effective working relationship begins to develop, thus putting the student at ease (Reeser, 1992). A prepracticum visit may also facilitate ongoing and positive communication among all practicum partners.

Monitor performance throughout practicum

The course instructor should be readily available to the students, preceptors, and disability services office personnel during the practicum placement. A check-in via email or a phone call with students and preceptors midway through the learning experience is helpful to ensure both the student and preceptor are having positive learning and teaching experiences. As with all students, if there is a possibility of student failure at this point, it is helpful for the course instructor to talk with both the preceptor and the student. The student may also find it helpful to contact his or her accessibility counselor to discuss possible strategies for continued success in the second half of the practicum.

Reflect on teaching and learning

As each student's learning experience is unique, it is helpful for the course instructor to meet with the student so that they can both reflect on the accommodation process and the effectiveness of the learning experience. Reflection is of benefit to students as they plan for other learning experiences and helps the educational institution and site preceptors improve the process and ultimately assist other students to equitably participate in practicum placements.

Input from the preceptor on the process including successes and challenges is also beneficial. Preceptors and course instructors have suggested students reflect on the practicum learning experience to identify facilitators and barriers to learning with accommodations in place and what could be done differently to facilitate learning in the next practicum placement. These reflections often benefit other students as well as the preceptor and course instructor when planning for future placements. In some cases, preceptors ask students to envision themselves working in the practicum environment as this reflection helps students identify the type of employment they may want to pursue after graduation.

Having the preceptor reflect on the teaching experience is also beneficial. Hirneth and Mackenzie (2004) as cited in Beckel (2012, p. 29 and 30) reported that preceptors found it difficult to obtain a balance between nondiscrimination and ensuring competent practice and also found it difficult to separate their roles (e.g., acting as the student's therapist). Some preceptors may also overcompensate for students with disabilities and not provide them with adequate opportunities to optimize learning experiences (Cooley & Salvaggio, 2002). The course instructor can help the preceptor review and reflect on the types of issues that can affect the teaching and learning experience.

Some questions to promote reflection include:

- For the student: How was the learning facilitated or compromised? Was the necessary information clearly communicated between all fieldwork partners? What would improve the accommodations process (in any stage of the PAPM)?
- For the practicum site: How was the teaching and learning facilitated or compromised? Was the necessary information clearly communicated between all fieldwork partners? What would improve the process for implementation of accommodations for future students?
- For the course instructor and other university faculty: Was the necessary information clearly communicated between all fieldwork partners? What would improve the accommodations process (in any stage of the PAPM)? What would facilitate learning for this student in the future?

Reflections should be documented for future reference by and for each fieldwork partner.

Considerations and strategies for successful provision of practicum accommodations

There are a number of useful considerations and strategies that will allow for the identification and implementation of different types of accommodations for students who participate in practicum placements. Individualized accommodations should be considered on a case-by-case basis by all practicum partners involved in the provision of the practicum placement. These accommodations can vary depending on a number of factors including the practicum's setting, the nature of the student's disability, and/ or the duration of the practicum placement, which can range from one day to many months. Regardless of the setting, duration of the assigned practicum placements, or the student's disability, the implementation of reasonable accommodations must occur.

Maintenance of academic and professional standards

Implementing reasonable student accommodations for practicum placements that comply with course, university, professional, and degree requirements can be challenging. All students must be held to the same academic and professional standards, regardless of disability. Students must acquire and demonstrate mastery of essential skills or other academic requirements of a student's program of study. For successful student accommodations in practicum learning, it is imperative that all partners think creatively, embrace flexibility, and remain open to consideration of different methods to meet course objectives and reach required competency levels. Standards cannot be lowered and requirements cannot be waived.

Course instructors should be familiar with university, professional, and regulatory college bodies that outline academic and professional standards and requirements in order to fully understand what accommodations are possible and reasonable while still maintaining the required standards. Consultation with university colleagues and professional peers who are familiar with these standards is helpful. The student requiring the accommodations should be educated about the required standards and competencies and be actively involved in the formulation of solutions for issues that arise. Questions for consideration and discussion of reasonable accommodations include: Is it possible for the student to demonstrate competency through detailed verbal or written instruction to an assistant rather than the student completing the physical action themselves? Does time need to be a factor in the demonstration of the competency? Can assistive devices be used to assist in the demonstration of competency or in competency development? Does demonstration of competency in a simulated environment/situation equal demonstration of competency in the actual environment/situation?

When designing a course that includes practicum placements, the course instructor should clearly outline the goals and objectives for the course, including learning objectives and outcomes specific to the practicum placements. In order to ensure accommodations are successfully implemented, flexibility regarding teaching methods, the learning environment, and how knowledge, skills, or attitudes will be evaluated needs to be considered. Additionally, professional programs should ensure all students are informed of the specific behavioral and clinical performance standards and how they will be evaluated for each practicum learning experience (Pardo, 1999).

In the development of practicum learning experiences, it is helpful for the course instructor to consider the following relevant principles of universal instructional design (UID): creating a welcoming environment where students feel empowered, defining essential components, providing clear expectations, providing feedback,

creating natural supports for learning, using varied instructional methods, and allowing students to demonstrate their learning achievements in multiple ways (Sharby
& Roush, 2008). Practicum learning that is designed with these principles in mind
allows students with a broad range of characteristics and abilities to fully participate
(Opitz & Block, 2006). A UID approach that is inclusive and sustainable unfortunately has not been universally utilized; therefore the individualized accommodation
approach is still more commonly used.

Time

The accommodations process for practicum placements can be a time-intensive process for all practicum partners. The type of student accommodations required will
dictate how much time it will take to construct the best possible learning experience
for the student and teaching experience for the preceptor.

Effective and timely communication with all partners requires regular meetings,
emails, and phone calls throughout the accommodations process. Taking the time to
have these meetings is essential and will result in a more satisfying and beneficial
experience for the student and preceptor. It is also helpful for all partners to anticipate
issues that might arise and brainstorm about possible solutions and plans of action
prior to the beginning of the experience so that issues are dealt with quickly and efficiently if and when they do arise (Reeser, 1992).

Detailed planning prior to the beginning of the learning experience helps to alleviate many potential difficulties (Beckel, 2012; Francis et al., 2007). Education of
the student and preceptor regarding course requirements and university expectations
should be done well in advance of the placement so there is ample time to answer
questions, clarify key points, and provide helpful resources. Whenever possible, education should be done in groups and/or be in the form of a recorded webcast in order
to use time as efficiently as possible.

The course instructor and student should also make full use of all available university resources and services to make the learning experience as successful as possible.
Faculty advisors and university counselors should be consulted before, during, and
after the learning experience as required. Obtaining necessary practicum site placement agreements and other documentation can be delegated to administrative staff or
teaching assistants.

Organizing and saving information and resources from year to year is extremely
beneficial for the course instructor. For example, making an inventory of which
practicum sites have been able to offer particular student accommodations is helpful
when assisting students in subsequent years (Dunn et al., 2008). A directory of contact
information for helpful university resources can also save time in the future. Having
copies of relevant legislation, policies, and regulations and a reference listing specific
and general answers and solutions to university, student, and practicum site questions
and challenges that have arisen can save time when similar issues arise again.

Course instructors should remember that the students themselves are great
resources for creative solutions to problems that may arise as many of them have had
dealt with similar situations throughout their lives and know their own strengths and
resources. Ensuring students are actively involved in the accommodations process in

practicum placements can save time and also provides a valuable practical experience for the student in dealing with any necessary accommodations in his or her future career. As mentioned previously, time spent reflecting on the whole PAPM process with the student and other partners is essential for continuous learning and quality improvement from year to year.

Equitable opportunities for all students

As noted previously, it is helpful to first assign practicum sites to students who require accommodations before assigning for the remainder of the class. This allows the student and course instructor to find placements that provide the necessary accommodations to the student. Unfortunately, students without accommodations may feel this process is unfair, and that they do not have the opportunity to receive an opportunity to be pre-placed in the most popular placement sites. In some cases, students with disabilities may actually have fewer choices of practicum sites due to the availability of sites where their unique learning needs can be adequately met. It is important to have clear, transparent process for student placements that can be discussed with all students upfront.

Student accommodations

Altered time

Flexible practicum hours such as the consideration of a part-time schedule or a shift in hours that the learning opportunity is typically offered can allow a student to equitably participate while still completing the requirements of the program. Some students may need to attend his or her practicum placements later in the day due to the nature of their disability, or leave early to attend medical appointments, and in some cases have their practicum placement extended to ensure required hours are met.

Other types of accommodations can include increased time allowed for daily preparation of activities and/or increased time allotted for documentation or program planning. Increased time spent with preceptors can allow students to obtain regular feedback on their progress. It is important to remember that if the student has to spend more time on tasks, this will take away from time clients, peers, preceptors, or other team members. This needs to be carefully monitored, since in some cases it may be difficult for some students to gain competence in time for evaluation.

In some cases, despite appropriate accommodations, students may require a leave of absence from the practicum placement. Depending on the program policies, students may be allowed to continue with other coursework and return to the practicum when appropriate.

Geographical location

There are several useful travel-related practicum placement accommodations that can be considered prior to and during learning opportunities. For example, prior to site selection, the geographical location of the site and the student's living accommodations should be considered so the student can be placed at a facility that best suits his or her learning needs and that is close to place of residence, medical facility, educational institution, etc.

Provision of one primary preceptor

Many practicum placement sites have more than one preceptor supervising students. Sharing two preceptors can be more challenging for any student, but even more so with some types of accommodations. Different preceptors often have different expectations and work styles, which can be difficult for students who require more structured opportunities. Having one primary preceptor may reduce student anxiety as the student will be evaluated by just one preceptor.

Homogenous practice area

It is helpful to offer flexible placement hours and start times for students who may require this as an accommodation, as long as the required number of hours is met. For example, patients with similar diagnoses may provide students with the opportunity to more quickly become familiar with certain conditions and treatment options.

Use of technology

Consistent with strategies that promote UID principles, technology can be used (Gupta et al., 2005), such as screenreaders, voice recognition software, voice amplifiers, specialized computing software, and/or the use of electronic documentation. Workspaces may also need to be ergonomically assessed so the student has the appropriate setup for desk, seating, and computer use. Additionally, it may be necessary to allow time for the student to alternate periods of standing and sitting and/or rest periods.

Considerations for national and international practicums

Students with disabilities cannot be denied national and international practicum placement opportunities. However, practicums outside of the home university catchment area require careful consideration by the student and the course instructor to ensure learning goals can be achieved and the student's accommodations can be provided. As often as possible, universities should use national and international practicum sites to ensure accessibility during the learning experience, which can be challenging for placements in some developing countries. Reflecting on all aspects of the PAPM can assist the student and course instructor in their discussions concerning these learning opportunities that are outside of the university catchment area. For example, will effective communication between all practicum partners be possible before and during the practicum? Does the course instructor have knowledge of the practicum site and fully understand the site's ability and willingness to provide the necessary student accommodations? Students themselves should consider their access to family, friends, health services, counselors, or specialists and how that might affect their performance during the learning experience.

Conclusion

The benefits and opportunities practicum placements offer learners are evident and must be accessible for all students. However, when practicum sites are not fully accessible, individualized student accommodations are required. The PAPM clearly outlines the considerations and steps involved in the successful provision of practicum placements for students with disabilities. The contextual factors (national, university, student, and practicum site) are integral components of the accommodation process and need to be considered during all stages of communication with practicum partners. It is this partnership that enables students to fully participate and optimize learning in their practicum placements. Challenges such as the maintenance of professional and academic standards and university faculty time constraints can be overcome when processes are followed and useful strategies are implemented. Although implementing student accommodations for practicum placements may appear complex, the benefits to all partners can be realized through greater awareness and understanding of inclusive education for all.

General principles of designing accessible learning environments in the sciences

Chapter Outline

Introduction   286
Practical spaces revisited   287
Differences between practical spaces and traditional science laboratories   287
Case study: Practical spaces in occupational and physical therapy   287
Case study: Archival spaces   289
The diversity of practical spaces in STEM education   289
Guiding principles for designing accessible learning environments in STEM   290
Overview of universal design principles   291
Flexibility   292
Dynamism   293
Collaboration   293
Fostering positive relationships   294
Does not contravene academic or professional rigor   294
Encompasses the many faces of a student in STEM   295
Conclusion   295

Why have I been successful?

Particularly in an environment where I am aware of so many stories of students who weren't successful?

What is it about my experience that stands out? What is it about the experiences of those others who were also successful in STEM?

And what lessons learned are there that we can draw from those positive experiences for others to take away?

In reality, those questions were the genesis of the endeavour behind this book, and were constant companions during its writing process. In more reflective moments, I found myself asking "what is success?" as well.

I know of students who consider "success" to be the passing of a given milestone marker – high school, college, university, graduate school, obtaining employment. I've witnessed the counterweight to that perspective, though: Having achieved those milestones, the student may no longer have a desire to strive for the next one.

Creating a Culture of Accessibility in the Sciences. DOI: http://dx.doi.org/10.1016/B978-0-12-804037-9.00025-5

"Success" could also be the capacity to compete on a level playing field with our nondisabled peers; to match their accomplishments or exceed them; or, to "out-scientist the scientists."

"Success" for me, I think, is a little more basic: It is the opportunity to chase my dream, to contribute in a meaningful way to the field I've come to love. Ultimately, that means that success is what we make it out to be.

I am also cautious of drawing too much from my own experiences. To recapture the specific circumstances surrounding and reasons for my success in others is too difficult; so, writing this book as a "how -to" manual was never something I wanted to do. I wanted to share the thought process that everyone involved in making me the world's first blind geneticist went through – how did we approach the problem at hand? How did we tackle the questions and challenges we faced? Irrespective of the specific circumstances faced by a given student, how to think through the problem rationally, creatively and effectively is a useful skill set and bit of background knowledge to have.

Introduction

Most literature on the accommodation of students with disabilities in the science lab setting has focused on chemistry education and/or accommodations at the secondary level. However, the concepts of accessibility and universal design (UD) are translatable across multiple contexts and disciplines. Anecdotal evidence from students and young scientists with disabilities in various disciplines suggests that accessibility of the science laboratory environment has a number of common themes. These include the importance of creativity in addressing academic accommodations, particularly with technology adaptations; a strong relationship, or "partnership," with faculty—either the course instructor/coordinator or the thesis supervisor; a flexible teaching approach; and, creativity in meeting the essential requirements for a course, program, and discipline.

Throughout this book, we have, where possible, utilized specific case studies and examples to illustrate accessibility and accommodation planning in science, technology, engineering, and mathematics (STEM) learning environments, with an emphasis on laboratory environments. An identified need for further research goes beyond describing barriers to science: research provides educators and disability service providers with resources and best practices that address all aspects of the accessibility of a science laboratory, as we have attempted to do here. However, no discussion can be fully exhaustive. In this chapter, we attempt to synthesize our experiences into a general discussion on designing and implementing accessible learning environments in STEM for students with disabilities. We offer a perspective and framework on how to approach designing such environments "from first principles" as it were, in ways that may be responsive to students' identified needs.

We will also return to the UD concept in this chapter, and ask a separate, yet related, question: What are the key principles in creating a culture of UD in the sciences, considering the sheer breadth of disciplines and research cultures reflected across the spectrum of training environments in postsecondary education?

Practical spaces revisited

In the previous chapters, we largely focused on the traditional science laboratory setting, i.e., the physical and biological sciences. In this chapter, we aim to expand the discussion of barriers for students with disabilities in the laboratory setting to include nontraditional lab environments. As in chapter 1: The landscape for students with disabilities in the sciences, we use the term "practical spaces" to refer to the nontraditional laboratory setting.

Practical spaces are defined as learning environments where students have the opportunity to engage in active learning and to demonstrate, through hands-on activities, the practical components of a given discipline. Practical spaces can be used to supplement classroom learning and to provide additional opportunities for students to work through practical scenarios typical of their respective disciplines. In the following case studies, we provide an overview of some of the common barriers and solutions to navigating the practical spaces students with disabilities face.

Differences between practical spaces and traditional science laboratories

Practical spaces can employ hands-on simulations using actors or volunteers. Practical spaces often are designed for the creation of material or some form of intellectual property, while teaching labs are more concerned with understanding principles behind processes. Practical spaces can be very modular and varied in design, but teaching labs generally conform to a more standard layout. Practical spaces can employ hands-on, human interaction-oriented simulations. Practical spaces often are designed for the creation of material or some form of intellectual property, while teaching labs are more concerned with understanding principles behind processes.

Case study: Practical spaces in occupational and physical therapy

In both occupational and physical therapy programs, there is a large amount of work students are required to complete in a laboratory setting. This labwork involves the integration of theory with practice, and provides an opportunity for the student to apply classroom knowledge to a practical laboratory setting (Barker & Stier, 2013). In these practical spaces, some examples of skills and core competencies students learn include how to assess patients, how to create splints, methods of cognitive assessment, and how to conduct patient exercises. This space is used to supplement the theoretical learning that occurs in the classroom and provide students with the opportunity to "practice" classroom discussion.

The barriers to students with disabilities in practical spaces are similar to those faced in the STEM lab setting. These barriers include challenges with the physical set-up of the laboratory, attitudinal barriers on the part of faculty, limited time to complete course requirements, and developing accommodations that will be feasible and realistic both during the program and in practice after graduation.

Given the physical nature of both occupational and physical therapy disciplines, students are required to lift/transfer patients, assist patients in moving from one area to another, and provide assistance in patient exercise. This is particularly challenging for students who have mobility or physical limitations. In these cases, one possible accommodation may be for the student with a disability to have an assistant who does the physical portion of the labwork under the student's specific direction. Although the student with a disability requires an accommodation to "physically" complete some of the laboratory tasks, it is important to note that all direction about the nature of the tasks comes from the student with the disability, and the assistant is simply there to carry out his or her instructions (Sukhai et al., 2014).

Engaging a paid or volunteer lab assistant provides the student with an alternate method of completing a task, but the ultimate learning outcome does not change. Faculty can be open to "different ways of doing" and have discussions with the student early in the course or program regarding the essential requirements of and learning outcomes for that course (Roberts, 2013). It is important and necessary for faculty and program coordinators to think critically about the course requirements and to determine which aspects of the course or program content are "essential require-ments" for completion of the curriculum (by the student) and which aspects can be accommodated using a lab assistant (Sukhai et al., 2014; Roberts, 2013).

Not unlike in the traditional science lab environment, faculty attitudes play a large role in determining a student's success in the program or discipline. Attitudes of fac-ulty and administrators can represent a stumbling block to students with disabilities, especially if a faculty member has a negative attitude around disability or around a student's ability to complete the lab component of the program. There can be an unwillingness to accommodate, or a subjective aversion to disability in the context of the learning environment, program, or discipline, and those students who have been labeled as such. These conditions create barriers to open communication with faculty, making the learning environment less than ideal.

Promoting faculty awareness of disability is one way to counteract this concern. There are two major strategies employed to aid faculty in developing disability-related knowledge. Some faculty simply have not encountered the need for accommo-dation. However, once they have had the opportunity to become informed they make the necessary changes to their program, teaching style, assessment of the student's knowledge of the course material, or his or her points of view. For other teaching staff, negative attitudes may be beyond any amelioration or transformation (Sukhai et al., 2014). Nevertheless, if a postsecondary institution has a culture of equity and acceptance, then even those who harbor their own prejudices tend to acquiesce to reasonable accommodations that do not compromise the student's ability to achieve the learning outcomes or the academic integrity of the program of study.

Case study: Archival spaces

Faculty who are instructing or facilitating courses with an archival or strong library-based component should be aware of the multiple barriers students with disabilities may experience in conducting archival research. While barriers in practical settings often potentiate one another, this occurs more in archive settings, where informational barriers and lack of awareness on the part of archival staff are uniquely connected. For example, staff who may not be familiar with how to convert material into an alternate format present an informational barrier for a student who requires said information for his or her research.

Archives can be challenging and confusing spaces to navigate and to retrieve information for those with no visual or physical disability. However, these challenges are amplified for those working in archival spaces who have a visual or physical disability. One of the major barriers for someone with a visual disability is locating and accessing information. Much of the printed information in archives is located in large books, some of which cannot be physically scanned or removed from storage. The information is in small print, and due to the fragile nature of the printed material, it is difficult to place these documents onto a flatbed scanner. While archive staff can provide general guidance on where to locate a specific topic, they may not be able to act as a reader and orally read information to a student who is blind. Additionally, archival staff are not in a position to digitize the information stored in the archives.

The diversity of practical spaces in STEM education

Many types of practical space and lab environments exist within postsecondary institutions, including typical basic science labs across a range of disciplines, engineering and applied sciences laboratories, computer labs, and labs within the context of some social and behavioral sciences (e.g., anthropology and psychology). Although we have used examples from the basic sciences, many of the concepts we have discussed so far are just as applicable to the spectrum of STEM-related learning environments in postsecondary education.

Our research and experience has led us to suggest that faculty and service providers working in STEM fields could apply the principles discussed throughout this book, and highlighted in this chapter, for a critical review of the accessibility of their relevant practical spaces. For example, engineering labs may have significant parallels to applied physics laboratories. Clinical lab learning environments in other allied healthcare disciplines can be considered analogous to OT and PT labs. Anthropology and archeology labs rely on many of the same tools and technologies found in biological sciences labs (e.g., dissections and microscopy) and in the archival setting (e.g., specimen preservation and storage). Biological psychology labs are analogous to more traditional biological science spaces. Food-science spaces can be considered as similar to chemistry environments, as can pharmaceutical laboratories. Computer and IT labs are heavily dependent on the accessibility of the technology and workstations, and practical spaces in other disciplines reliant on computing are likewise dependent on the principles of IT accessibility (cf. Berliss, 1991).

Guiding principles for designing accessible learning environments in STEM

Based on a synthesis of our learning, as presented throughout this book, we recommend a "thought rubric" for faculty, staff, and services providers in understanding, evaluating, and developing solutions for accessibility concerns in practical spaces across disciplines. This rubric is framed as, and poses, a series of questions for faculty, staff, and services providers to ask in understanding both barriers and potential solutions to accessibility concerns for students with disabilities, and focuses on the principles of faculty–student engagement, inclusive teaching practices, UD, and essential requirements.

Questions to ask in order to understand the barriers inherent in a practical space include:

- What are the essential requirements of the course, program, or discipline?
- Through a rational deconstruction of the components of the practical space (including "walkthroughs" of the learning process with the student), what are the key issues that are likely to arise for the student?
- What tasks might the student have most difficulty with, and how do they relate to the essential requirements of the program, discipline, or course?
- What barriers (physical, technological, attitudinal, other) to the participation of students with disabilities exist in the context of the practical space?
- Are there learning outcomes the student must demonstrate unaided? If so, what are those outcomes?
- What outcomes can students demonstrate with the appropriate assistance or accommodation?
- Is the learning material being presented in the practical space meeting the principles of UD for learning?
- What systemic barriers and system issues may exacerbate these considerations? Within the department? Within the university or college? Within the field or discipline itself?

Questions to ask in order to identify potential solutions for the student include:

- What lessons by analogy can be applied from other disciplines, where accessible learning environments might be better understood?
- Will a laboratory/technical assistant be a feasible or appropriate accommodation/solution?
- What mainstream technology can be adapted as an accommodation aid for the student?
- What potential solutions can be evolved from experience that may be workable for the student, without compromising the essential requirements of the discipline?
- What alternative learning strategies can be employed in order to achieve the course, program, or discipline objectives?

As the needs of individual students with disabilities are different, it is important to remember the importance of using this thought process with all students with disabilities. While accommodations can be translated from one student to the next, it is important to remember that the effectiveness of their application will be dependent on the individual student.

Carrying out this thought process enough times leads to learning environments that converge on UD principles. Rather than suggest the specifics of those environments, and again recognizing the sheer breadth of possibilities in STEM training and the relative impossibility of a one-size-fits-all approach, we offer for the reader's consideration our learning around key principles that form the foundation for the adoption of UD principles in STEM education.

Overview of universal design principles

In STEM education (as with education broadly), individualized accommodation (also known as personalized accommodation) of students with disabilities is the norm. This approach involves the provision of supports and services based on the abilities and needs of each person. While individualized accommodation may be the current best practice, it is often costly, time-consuming (Pavri, 2010), and retroactive (Harrison, 2006). Furthermore, in STEM learning environments in particular, individualized accommodation is often hampered by a student's not feeling comfortable disclosing his or her disability, a lack of collaboration among students, disability services staff, and faculty, and inadequate knowledge of previous precedents.

A lack of accommodation can then impact the learning and attainment of students who depend on supports and services to sustain their studies. Moreover, personalized accommodation relies on students disclosing their respective accommodation needs, typically through provision of a medical diagnosis of their disabilities. Disclosure can be a difficult and complex decision for many students who fear the stigma attached to a disability label (see chapter 3: Barriers faced by students with disabilities in science laboratory and practical space settings). The dilemma surrounding disability disclosure can be particularly prominent in STEM education, where competence and autonomy are not only highly regarded but also traits students are expected to possess, especially at the graduate level (Council of Ontario Universities, n.d.).

Concerns around disclosure may be heightened for graduate students with invisible disabilities compared to graduate students with visible disabilities because they have to determine whether or not to reveal their disabilities (Côté, 2009).

UD is geared toward creating barrier-free environments for everyone and consequently is often promoted as a panacea to the challenges of individualized accommodation. As we have seen previously, UD is intended to ensure products and environments are "usable by all people, to the greatest extent possible, without the need for adaptation or specialized design" (The Center for Universal Design, 1997). Derived from UD are universal design for learning, also known as UDL (Rose & Meyer, 2002), which is focused on ways to display knowledge and skill acquisition; and, universal design of instruction (UDI), which is described as "an approach to course design that seeks to create an appropriate learning environment for all students, including those with disabilities" (Shaw, 2011, p. 21).

While research has been done on the benefits of UDL and UDI for undergraduate students, the meaning and impact of UD in more specialized settings such as graduate education and STEM training (where much of the student's learning takes place outside the traditional classroom environment) has yet to be considered beyond the traditional classroom setting. In order for UD to be relevant, responsive, and beneficial to students with disabilities and all students more broadly, we must examine the principles of effective UD in STEM education.

This section will outline several principles that may serve as guidelines when universally designing STEM education environments. The principles discussed here evolved from an examination of the perspectives of students with disabilities about the factors that contributed to their success. The principles were also derived from our discussions with faculty, professionals working in student services, and other stakeholders who assist students with disabilities. While the focus here is on students with disabilities in STEM, it is vital to recognize that UD can enhance the university experience for all students, not only those with disabilities. It should also be noted that the UD principles we propose herein do not exist in tension with, or supersede, personalized accommodation but rather it is hoped that their identification will serve to facilitate enhanced supports for students with disabilities.

The principles to be discussed are as follows:

1. Flexibility: Relates to the capacity of a STEM education environment to respond to the diverse abilities and needs of students with disabilities.
2. Dynamism: Focuses on the ability of STEM programs and environments to adapt to students' changing needs and circumstances, whether they be academic or personal in nature.
3. Collaboration: Stakeholders working together and communicating openly with one another to ensure students are well-supported and their needs met.
4. Fostering positive relationships: Relates to interactions between peers as well as to interactions between faculty and students.
5. Does not contravene academic rigor: Pertains to the balance that must be achieved between meeting the needs of students while not compromising the integrity of a program or institution.
6. Encompasses the many faces of a student in STEM: Recognizes the ways in which STEM education is distinct and takes into consideration the myriad responsibilities students adopt in STEM training as part of their education.

Flexibility

Flexibility relates to the capacity of a STEM education environment to respond to the diverse abilities and needs of students with disabilities. Personalized accommodation in STEM education, although intended to be specific to the individual, can in many instances consist of a limited repertoire of generic supports. These supports may be dependent on the availability of resources and make assumptions about the individual's needs based on what is traditionally provided to students with disabilities in other circumstances (e.g., the classroom setting), such as being afforded additional time to complete exams. On the other hand, UD ensures that supports and services are embedded in the environment, proactively, ideally *before* students encounter struggles, potentially reducing the need for accommodation. A universally designed STEM education environment is more flexible and capable of responding to the diverse needs of different students with disabilities, because many of the frameworks for response are already in place, or have already been thought through by faculty and disability services staff. In this way, a universally accessible STEM education environment will recognize that a student's program, needs, and circumstances can evolve naturally, and not rely heavily on potentially ineffective, generic supports if and when challenges arise. Moreover, while the behaviors, needs, and expectations of students may be similar in some respects, universally designed environments strive to encompass the diversity of program requirements students must meet and the roles they must fulfill throughout STEM training.

Dynamism

Individualized accommodations tend to be provided reactively (Morgan & Houghton, 2011). Additionally, the services and supports available to students can vary greatly in their quality and scope from one area and even campus to the next (Stodden & Conway, 2003). It is also important to note that some students' disabilities may be unpredictable in nature (Brown, 2008) with regard to the ways in which they affect students' health, learning, engagement, and daily living. For example, a student may find it much harder to maintain consistent attendance in the winter than the summer months. Additional needs or challenges can emerge as students grow more immersed in their programs and/or students find that previously utilized modes of coping and management are ineffective. Some students could be impacted, either positively or negatively, if their disabilities are progressive in nature, the work in which they are engaged changes or advances (such as from taking courses to thesis writing), or new technology becomes available for use in their respective programs. McEwan and Downie (2013) suggest that students with mental health-related disabilities do not respond favorably to a self-advocacy-based model of support. A student's disability, program, and other circumstances will change over time—and in ways that may or may not be cyclical. As with the concept of flexibility, this concept encourages constant dialog among the student, services provider, and faculty, in a manner that is sensitive to students' concerns about disclosure and stigma (e.g., for students with mental health disabilities). This concept also encourages rapid recognition of when a specific set of supports may not any longer be appropriate and requires modification.

Collaboration

While individualized accommodation typically centers on discussions between a student and a disability support advisor, a universally designed approach might favor teamwork. This could consist of the student, his or her supervisor and professors, anyone whom the student wishes to bring in, and anyone who needs to be involved on the individual's behalf. It is true that, particularly given the complexity of STEM training, the engagement of multiple stakeholders may be more conducive to fully understanding the student's needs and determining how best to address them. As a result, universally designed STEM learning environments would be collaborative, continually evolving to meet the needs of students and their programs. This involves recognition of the fact that students' needs may vary depending on what is being asked of them at different points in time and how their personal circumstances develop and evolve.

Fostering positive relationships

Although helpful, requesting personalized accommodations can cause students to feel stigmatized and sometimes isolated. These accommodations may also create barriers to the establishment of strong peer and faculty–student relationships. There is the potential for students with disabilities to feel isolated from peers because they require accommodations that other students do not. Faculty may also develop misconceptions of a student when accommodations are being provided (Burgstahler, 2003), before they have really had an opportunity to become acquainted with that individual's strengths and challenges beyond what is written on paper. A universally designed STEM education environment would cater to students' differences by allowing them to demonstrate learning and knowledge and participate in the environment in ways that align with their personal strengths. It would also ensure they are naturally well supported without drawing unnecessary attention to their needs. This may help students with disabilities feel more comfortable in social interactions and also allow faculty to get to know students as unique individuals rather than their disabilities.

Does not contravene academic or professional rigor

Academic rigor and professional competence are highlighted by higher education institutions as two of the cornerstones of high-quality graduate programs and schools (e.g., Ryerson University's master of professional communication program). In principle, admissions criteria and program requirements are designed to safeguard academic and professional rigor. In practice, these strict admissions requirements may be a barrier to entry for many prospective students (some with disabilities) because their skills and experiences do not fit the traditional mold of what constitutes a capable student (Cross, 1981). Additionally, students who are admitted to programs may be stymied by rigid program requirements that do not take into account the difficulties they encounter in satisfying such requirements because of their disabilities. A universally designed STEM education environment would maintain the academic and professional rigor of these programs but recognize that this can be demonstrated and fulfilled in different ways. Furthermore, UD could support students with disabilities to satisfy program requirements by preserving the overarching competencies associated with these requirements but allowing students to tackle them in a way that reflects their different abilities and strengths.

Encompasses the many faces of a student in STEM

Individualized accommodations typically focus on campus-based instruction, which involves assignment and exam-based forms of assessments. However, depending on a student's program, he or she may wish (or be required) to complete course and/or labwork, serve as a teaching assistant, undertake research, complete fieldwork or a practicum, and participate in professional development opportunities (for a more wholesome discussion, see chapter 15: The student in a leadership, mentorship, and supervision role). Students may also engage in volunteerism or extracurricular service, such as student associations or academic councils (which may consist of both on-campus and off-campus duties), in order to:

1. Contribute to the betterment of their respective universities or the wider community;
2. Augment their skills and knowledgebase;
3. Improve their preparedness for future employment; and,
4. Increase their competiveness when applying for research grants and scholarships.

Additionally, students at the PhD level must take and pass a candidacy exam before they are permitted to conduct their thesis research, a stipulation that does not exist at any other level. It is also necessary to recognize that being a graduate student often involves traveling for conferences and presentations. Consequently, a universally designed environment would be multifaceted, with people being able to take advantage of the supports and services that are embedded within the various areas that comprise their programs and lives as students. This would allow students with disabilities in STEM the level of flexibility they need to be successful while not having to compromise their responsibilities, quality of life, or personal standards of achievement.

Conclusion

Although complex, it is crucial that we consider the nature of UD in STEM education when designing courses, determining program requirements, recruiting students, and designing supports and services. Traditional modes of accommodation, while well intentioned, are insufficient to fully meet the needs of students with disabilities in STEM disciplines. In this way, the purpose of universally designed STEM education environments is neither to ignore nor to force disclosure of differences. On the contrary, its goal is to foster an overall culture in which students feel comfortable disclosing their differences if they wish to do so, without fear of recrimination or misjudgement. It must also be stressed that UD does not preclude the provision of individualized accommodation if needed. In fact, individualized accommodations may still be needed even with environments being universally designed and better equipped for students to complete their degrees. Thus, in order to be successful, the development of universally designed STEM education environments must be multi-layered, paralleling the nuances of student life.

Finally, it is clear that for universally designed STEM education environments to truly be universal, they must not only be usable by all students but also serve as the product of continuous, collaboratively oriented, in-depth discussion and debate between all education stakeholders. While certainly challenging to implement, this kind of teamwork highlights not only the position of students in STEM training but also the value and importance of voices in STEM education coming together to fuel positive change at the level of policy and practice.

Conclusion: STEM and disability—a vision for the future

In the preceding pages, we attempted to provide an overview of the landscape for students with disabilities pursuing science, technology, engineering, and mathematics (STEM). Based on our findings, and the current literature to date, a small number of students pursue STEM at a postsecondary level, and indeed, an even fewer number of students pursue STEM postgraduation. Not surprisingly, students discontinue their interest in STEM due to innumerable barriers faced throughout the elementary, secondary, and even postsecondary education levels. These barriers are well-documented and include attitudes on the part of educators and family; lack of knowledge surrounding how to instruct students with disabilities in STEM courses; limited resources to support educators around how to instruct students; and, physical, technological, pedagogical, and cultural challenges to accessing a science laboratory.

The ability to access science education and training in STEM is a fundamental human rights issue in the context of the international education landscape, as basic numeracy and scientific literacy are as required as other forms of literacy for success and full participation in all levels of today's global society, and effective access to the workforce (STEM-related or otherwise).

More training of educators, accommodation specialists, employers, students and caregivers is required around STEM and disability, which will build a culture of inclusion and may lead to increased numbers of students going into STEM fields. Students are not being encouraged (by educators, guidance counselors, and support staff, in particular) about STEM, and they are especially not being talked with about STEM in a positive manner, with a focus on higher education and ultimately career options. While barriers to accessing STEM education and therefore careers in STEM fields have been delineated from both the physical and attitudinal perspective, broader discussions around creative solutions to these barriers, and truly fostering a culture of accessibility in the sciences, are still lacking.

While it is easy to focus solely on the physical barriers and access issues within the laboratory setting, these are merely one facet of the overall picture. It is critical that students who do wish to pursue STEM-related courses, programs, and disciplines develop strong self-advocacy skills and have a strong sense of their disability-related accommodations. This will allow them to more accurately speak to their educational needs and how to accommodate them.

There is currently little in the way of role models for students with disabilities in STEM, and this presents challenges when students need to seek out information that would better enable them to advocate for their own learning. Students with disabilities seldom have access to mentors with or without disabilities in STEM fields who are successful in careers that they might otherwise have thought impossible for themselves.

Given the documented barriers students with disabilities face when entering STEM, it is not surprising that those who are successful in entering STEM careers face a series of conceptual glass ceilings in the STEM training pipeline—at the entry to the undergraduate degree, at the entry to graduate and postdoctoral training, throughout their integration into the STEM learning environment, and at the point of employment. It is critical for students to have access to individuals who can champion their success and who are *willing to work with students to problem-solve*. Thus it is also strongly recommended that students have access to mentors or champions who can guide them through the process of navigating the myriad barriers in STEM and can work with students, educators, and professionals in the scientific community to deconstruct the glass ceilings in the STEM training pipeline for students with disabilities.

It is critical for educators, family, and services providers to not make assumptions based on past experience, or lack thereof, and instead, to work closely with students to understand their needs and the interface between the course, program, and discipline, and students' accommodation needs. Self-advocacy and appropriate disclosure are ultimately the responsibility of a postsecondary student. The student knows his or her disability and needs best, and is in the best position to speak to accommodation needs, with the supports and tools in place to help him or her do so. Some students, whether at an undergraduate or graduate level, choose to disclose their disability publicly in order to promote greater inclusivity for both themselves and other students with disabilities. The decision to disclose is highly personal and is therefore undertaken for varied reasons. Students who proceed down this more publicly oriented path become advocates, communicators, and trailblazers for future generations of students with disabilities in STEM.

Although faculty should provide appropriate accommodations, course, program, and discipline requirements should never be altered when assessing a student. Doing so can compromise the rigor of the program both for the student with the disability and for other students both with and without disabilities. Such an understanding, and the willingness to work with the student and keep an open mind, will foster student success.

Creating and maintaining a culture of accessibility significantly impacts the student's experience, and is critical in order to move beyond the idea that access is merely a physical construct. Thinking about access more holistically is essential in defining a student's ability to participate fully in science labs at the graduate and undergraduate levels. These include teaching practices, the relationship between the student and faculty members, and the accessibility of content-delivery methods and lab equipment. Furthermore, a focus on the physical accessibility of the lab addresses barriers faced by students with physical and sensory disabilities, while potentially minimizing the challenges faced by students with "invisible," mental health, learning, or cognitive disabilities.

Faculty, educators, students, and disability services staff will all find critical learnings throughout each chapter of this book. While the source material is the same for all audiences, the specific learned outcomes are unique to each group in different situations. We argue for creativity, cooperation, and collaboration among all target audiences in order to create a truly accessible, universally designed learning environment for all phases of science, technology, engineering, and mathematics training. Conditions and circumstances naturally evolve, based on a person's disability, accommodation needs, program, field of study, and research. It is thus critical to meet such changing circumstances with flexibility and responsiveness.

Students, faculty, and disability services staff all need to be open-minded, each in their own way, with respect to accessibility and disability in the sciences. We argue for the use of this book as a resource to demonstrate to students, staff, and faculty that they are not alone in attempting to evolve solutions to accessibility issues in the sciences; many precedents for thought and problem-solving exist in the field. There are also many myths and misconceptions surrounding the accessibility of STEM education and the inclusion of persons with disabilities in the sciences, which do not hold under rigorous and critical evaluation. Finally, and most globally, we argue for the existence of champions in the field—successful students, engaged faculty—who will advocate for greater inclusion of students with disabilities in STEM, and for a more universally designed science learning environment. By describing the diverse and unique learning environments open to students in today's changing learning landscape as well as the different faculty interactions and accommodation solutions possible in STEM education at the postsecondary level, it is hoped that students themselves will be better prepared to navigate their learning environments and educational journey.

In today's evolving society and labor market, it is critical that all students, with and without disabilities, are equipped to handle the changing economic landscape. This means at the very least, having a cursory understanding of STEM, and at best, the opportunity to pursue STEM fields if the student desires. Excluding this population of students from STEM opportunities is contributing to an untapped potential of qualified leaders in the scientific community. To engage more students in STEM disciplines, it is our hope that faculty, services providers, educators, peers, and families will provide students with disabilities with the tools necessary to contribute to a diverse and changing workforce and that students themselves will position themselves to best succeed with these tools and resources.

Bibliography

Abdulmohsen, A. (2010). Simulation-based medical teaching and learning. *Journal of Family and Community Medicine*, *17*(1), 35–40.

Alberta Human Rights Commission (2009). Bona fide occupational requirements. Retrieved from: <http://www.albertahumanrights.ab.ca/employment/employer_info/employment_contract/bfor.asp>.

Allan, T. (2014). Simulation as a learning strategy: Supporting undergraduate nursing students with disabilities. *Journal of clinical nursing*, *23*, 402–409.

Allen-Ramdial, S., & Campbell, A. (2014). Reimagining the pipeline: Advancing STEM diversity, persistence, and success. *Bioscience*, *64*(7), 612–618.

Al-Salehi, S. M., Al-Hifthy, E. H., & Ghaziuddin, M. (2009). Autism in Saudi Arabia: Presentation, clinical correlates and comorbidity. *Transcult Psychiatry*, *46*(2), 340–347.

Americans with Disabilities Act of 1990, as Amended, Pub. L. No. 110-325. (2008). Retrieved from <http://www.ada.gov/pubs/adastatute08.htm>.

Azzopardi, T., Johnson, A., Phillips, K., Dickson, C., Hengstberger-Sims, C. Goldsmith, M., et al. (2010). Developing culturally responsive approaches with Southeast Asian American families experiencing developmental disabilities. *Pediatrics*, *126*(Suppl 3), S146–S150.

Badenhorst, C., Moloney, C., Rosales, J., Dyer, J., & Ru, L. (2015). Beyond deficit: Graduate student research-writing pedagogies. *Teaching in Higher Education*, *20*(1), 1–11.

Balmer, D., Master, C., Richards, B., & Giardina, A. (2011). Implicit versus explicit curricula in general pediatrics education: Is there a convergence? *Pediatrics*, *124*(2), e347–e354.

Baker, R. S. J. D., Moore, G., Wagner, A., Kalka, J., Karabinos, M., Ashe, C., et al. (2011). The dynamics between student affect and behavior occurring outside of educational software. *Proceedings of the 4th bi-annual international conference on affective computing and intelligent interaction*, 14–24.

Barker, D., & Stier, J. (2013). *Consideration of student accessibility when teaching outside the classroom*. University of Toronto.

Beckel, C. (2012). *Clinical education accommodations for physical therapist students with disabilities*. Doctoral dissertation, Saint Louis University.

Berliss, J. (1991). *Checklists for implementing accessibility in computer laboratories at colleges and universities*. Madison, WI: University of Wisconsin. Retrieved from: <http://trace.wisc.edu/docs/accessible_labs/campus.htm>.

Blake-Drucker Architects. (2009). Accessible laboratory: beyond ADA – guidelines for universal access.

Brock University (2012). Seven principles of universal instructional design.

Brown, V. (2008). *Experience with support services of graduate students with disabilities studying at a distance: A case study*. Alberta: Master of Distance Education, Athabasca University.

Burgstahler, S. (2003). Accommodating students with disabilities: Professional development needs of faculty. In C.M. Wehlburg, & S. Chadwick-Blossey, (Eds.) *To improve the academy: Vol. 21* (pp. 179–195).

Burgstahler, S. (2012). *Making science labs accessible to students with disabilities*. Retrieved from <http://www.washington.edu>.

Burgstahler, S., & Cronheim, D. (2001). Supporting peer-peer and mentor-protégé relationships on the Internet. *Journal of Research on Technology in Education, 34*(1), 59–74.

Burgstahler, S. E. (2008). Universal *design in higher education*. In S. E. Burgstahler & R. C. Cory (Eds.), *Universal design in higher education: From principles to practice* (pp. 3–20). Cambridge, MA: Harvard Education Press.

Campaign for Science and Engineering. (2014). Improving diversity in STEM. Retrieved from: <http://www.sciencecampaign.org.uk/resource/ImprovingDiversityinSTEM2014.html>, October 16, 2016.

Cavell, T., DuBois, D., Karcher, M., Keller, T., & Rhodes, J. (2009). *Strengthening mentoring opportunities for at-risk youth*. Retrieved from <http://www.mentoring.org/downloads/mentoring_1233.pdf>.

Center for Applied Special Technology. (2011a). *CAST Timeline: One mission, many innovations, 1984-2010. (Web Page)*. Wakefield, MA: CAST.

Center for Applied Special Technology. (2011b). *UDL Questions and Answers (Web Page)*. Wakefield, MA: CAST.

Center for Applied Special Technology. (2011c). *What is universal design for learning? (Web Page)*. Wakefield, MA: CAST.

Center for Universal Design. (1997). *The principles of universal design*. North Carolina State University, Centre for Universal Design.

Chambers, T., Sukhai, M., & Bolton, M. (2011). *Assessment of debt load and financial barriers affecting students with disabilities in Canadian postsecondary education – Ontario report*. Higher Education Quality Council of Ontario. Retrieved from: <http://odenetwork.com/assessment-of-debt-load-and-financial-barriers-affecting-students-with-disabilities-in-canadian-postsecondary-education-ontario-report-for-the-higher-education-quality-council-of-ontario/>.

Chelberg, G., Harbour, W., & Juarez, R.L. (1998). Accessing student life: Steps to improve the campus climate for disabled students.

Chickering, A., & Gamson, Z. (1987). Seven principles for good practice in undergraduate education. *AAHE Bulletin, 38*(7), 3–7.

Cliffe, E. (2009). Accessibility of mathematical resources: The technology gap. *MSOR Connections, 9*(4), 37–42.

Cook, L., Rumrill, P. D., & Tankersley, M. (2009). Priorities and understanding of faculty members regarding college students with disabilities. *International Journal of Teaching and Learning in Higher Education, 21*(1), 84–96.

Cooley, B., & Salvaggio, R. (2002). Ditching the 'dis' in disability: Supervising students who have a disability. *Australian Social Work, 55*(1), 50–59.

Costa, M. M., & Gatz, M. (1992). Determination of authorship credit in published dissertations. *Psychological Science, 3*, 354–357.

Curry, C., Cohen, L., & Lightbody, N. (2006). Universal design in science learning. *Science Teacher, 73*(3), 32–37.

D'Angelo, C., Rutstein, D., Harris, C., Bernard, R., Borokhovski, E., & Haertel, G. (2014). Simulations for STEM learning: Systematic review and meta-analysis.

Dawson, T. (Ed.) (2004). Universal instructional design: Creating an accessible curriculum. *Teaching and learning services and accessability services* University of Toronto Scarborough.

Department of Justice. (1982). *Canadian Charter of Rights and Freedoms*. Retrieved from <http://laws-lois.justice.gc.ca/eng/Const/page-15.html>.

Dornan, T., & Bundy, C. (2004). What can experience add to early medical education? Consensus survey. *BMJ, 329*(7470), 834.

Doyle, T. (2014). Checklist for making science labs accessible for students with disabilities. Retrieved from: <http://www.accessiblecampus.ca/wp-content/uploads/2014/06/05.-Checklist-for-Making-Science-Labs-Accessible-for-Students-with-Disabilities.pdf>, October 16, 2016.

Dunn, P., Hanes, R., Hardie, S., Leslie, D., & MacDonald, J. (2008). Best practices in promoting disability inclusion within Canadian schools of social work. *Disability Studies Quarterly, 28*(1).

Dupler, A. E., Allen, C., Maheady, D. C., Fleming, S. E., & Allen, M. (2012). Leveling the playing field for nursing students with disabilities: Implications of the amendments to the Americans with Disabilities Act. *Journal of Nursing Education, 51*(3), 140–144.

Elander, J., Harrington, K., Horton, L., Robinson, H., & Reddy, P. (2006). Complex skills and academic writing: A review of evidence about the types of learning required to meet core assessment criteria. *Assessment & Evaluation in Higher Education, 31*(1), 71–90.

Equality Act. (2010). Retrieved from <http://www.legislation.gov.uk/ukpga/2010/15/contents>.

Fichten, C. S., Goodrick, G., Amsel, R., & Libman, E. (1996). [Original article and title are in Japanese]. Teaching college students with disabilities: A guide for professors. In Y. Tomiyasu, R. Komatsu & T. Koyazu (Eds.), *Support for university students with disabilities: A new feature of universities* (pp. 233–323). Tokyo: Keio University Press.

Francis, N. J., Salzman, A., Polomsky, D., & Huffman, E. (2007). Accommodations for a student with a physical disability in a professional physical therapist education program. *Journal of Physical Therapy Education, 21*(2), 60.

Galvaan, R. (2012). Occupational Choice: The Significance of Socio-Economic and Political Factors. In G. E. Whiteford & C. Hocking (Eds.), *Occupational Science: Society, Inclusion, Participation.* Oxford, UK: Wiley-Blackwell. http://dx.doi.org/10.1002/9781118281581.ch11.

Gerholm, T. (1990). On tacit knowledge in academia. *European Journal of Education, 25*(3), 263–271.

Giesen, J. M., Cavenaugh, B. S., & McDonnall, M. C. (2012). Academic supports, cognitive disability and mathematics multilevel modelling approach. *International Journal of Special Education, 27,* 1.

Gill, C. (1994). *Two models of disability.* Chicago (IL): University of Chicago, Chicago Institute of Disability.

Girves, J. E., & Wemmerus, V. (1988). Developing models of graduate student degree progress. *Journal of Higher Education, 59*(2), 163–189.

Golde, C. M. (2005). The role of the department and discipline in doctoral student attrition: Lessons from four departments. *Journal of Higher Education, 76*(6), 669–700.

Graham-Smith, S., & Lafayette, S. (2004). Quality disability support for promoting belonging and academic success within the college community. *College Student Journal, 38*(1), 90.

Greene, M. J. (2014). *The hardest part is through: Support services and graduate student persistence in the social sciences and humanities disciplines (Doctoral Thesis).* Canada: Memorial University of Newfoundland.

Groce, N. E. (1999). Disability in cross-cultural perspective: Rethinking disability. *The Lancet, 354,* 756–757.

Gupta, J., Gelpi, T., & Sain, S. (2005). Reasonable accommodations and essential job functions in academic and practice settings. *OT Practice, 10,* 15.

Hall, T., Healey, M., & Harrison, M. (2002). Fieldwork and disabled students: Discourses of exclusion and inclusion. *Transactions of the Institute of British Geographers, 27*(2), 213–231.

Harrison, E. G. (2006). Working with faculty toward universally designed instruction: The process of dynamic course design. *Journal of Postsecondary Education and Disability, 19*(2), 152–162.

Hartman, D. J. (1990). Undergraduate research experience as preparation for graduate school. *The American Sociologist,, 21*(2), 179–188.

Harvey, J. C., & Katz, C. (1984). *If I'm so successful, why do I feel like a fake?* New York, NY: Random House.

Healey, M., Jenkins, A., Leach, J., & Roberts, C. (2001). *Issues providing learning support for disabled students undertaking fieldwork and related activities.* Cheltenham and Gloucester College of Higher Education, Geography Discipline Network (GDN).

Heidari, F. (1996). Laboratory barriers in science, engineering, and mathematics for students with disabilities. A study conducted under a grant from the Regional Alliance for Science, Engineering, and Mathematics at New Mexico State University. (ERIC Documentation Reproduction Service, No. ED 397 583).

Herrera, C., Grossman, J.B., Kauh, T.J., Feldman, A.F., McMaken, J., & Jucovy L.Z. (2007). *Making a difference in schools: The Big Brothers Big Sisters school-based mentoring impact study.* Public/Private Ventures. Retrieved from <http://www.ppv.org/ppv/publications/assets/220_publication.pdf>.

Hewlett, M.G., & Burnett, A.N. (2006). *Text-to-Speech software and webCT: Issues of compatibility.* Retrieved from <http://web.unbc.ca/wccce05/Text-to-Speech_Software.doc>.

Higbee, J. L. (2008). The faculty perspective: Implementation of universal design in a first–year class. In S. E. Burgstahler & R. C. Cory (Eds.), *Universal design in higher education: From principles to practice* (pp. 61–72). Cambridge, MA: Harvard Education Press.

Hilliard, L., Dunston, P., McGlothin, J., & Duerstock, B. (2011). *Designing beyond the ADA— creating an accessible research lab for students and scientists with physical disabilities.* Lafayette, IN: Institute for Accessible Science, Purdue University.

Hirneth, M., & Mackenzie, L. (2004). The practice education of occupational therapy students with disabilities: Practice educators' perspectives. *The British Journal of Occupational Therapy, 67*(9), 396–403.

Howlin, F., Halligan, P., & O'Toole, S. (2014). Development and implementation of a clinical needs assessment to support nursing and midwifery students with a disability in clinical practice: Part 1. *Nurse Education in Practice, 14*(5), 557–564.

Hughes, M., Milne, V., McCall, A., & Pepper, S. (2010). *Supporting students with Asperger's syndrome: A physical sciences practice guide.* United Kingdom: Higher Education Academic, UK Physical Sciences Centre.

Isaacson, M. D., Srinivasan, S., & Lloyd, L. L. (2010). Development of an algorithm for improving quality and information processing capacity of MathSpeak synthetic speech renderings. *Disability & Rehabilitation: Assistive Technology, 5*(2), 83–93. http://dx.doi.org/10.3109/17483100903387226.

Jacks, P., Chubin, D. E., Porter, A. L., & Connolly, T. (1983). The ABCs of ABDs: A study of incomplete doctorates. *Improving College and University Teaching, 31*(2), 74–81.

Jekielek, S., Moore, K. A., & Hair, E. C. (2002). *Mentoring programs and youth development: A synthesis.* Washington, DC: Child Trends. Retrieved from <http://www.mentorwalk.org/documents/mentoring-synthesis.pdf>.

Johnson, A. (2006). *Students with disabilities in postsecondary education: Barriers to success and implications for professionals.* VISTAS online. Retrieved from: <http://counselingoutfitters.com/Johnson.htm>.

Johnson, D. (2000). Enhancing out-of-class opportunities for students with disabilities. *New Directions for Student Services, 2000*(91), 41–53.

Johnston, N., & Doyle, T. (2009). *Inclusive teaching: Perspectives of students with disabilities [Survey]*. Scarborough, ON M1C: University of Toronto Scarborough.

Johnston, N., & Doyle, T. (2011). Inclusive teaching: Student perspectives. *Open Words: Access and English Studies Journal, 5*(1), 53–60.

Kornblau, B. L. (1995). Fieldwork education and students with disabilities: Enter the Americans with Disabilities Act. *American Journal of Occupational Therapy, 49*(2), 139–145.

Kuh, G. D., Laird, T. F. N., & Umbach, P. D. (2004). Aligning faculty activities & student behavior: Realizing the promise of greater expectations. *Liberal Education, 90*(4), 24–31.

Langley-Turnbaugh, S. J., Murphy, K., & Levine, E. (2004). Accommodating students with disabilities in soil science activities. *Journal of Natural Resources and Life Sciences Education, 33*, 155–160.

Logan, J. (2009). *Learning disabilities: A guide for faculty at Ontario universities*. Ottawa, Canada: Council of Ontario Universities. Available from <http://www.cou.on.ca/publications/academic-colleague-papers/pdfs/learning-disabilities>.

LoSciuto, L., Rajala, A. K., Townsend, T. N., & Taylor, A. S. (1996). An outcome evaluation of across ages: An intergenerational mentoring approach to drug prevention. *Journal of Adolescent Research, 11*(1), 116–129.

Lovitts, B. E. (2001). *Leaving the Ivory Tower: The causes and consequences of departure from doctoral study*. New York, NY: Rowman & Littlefield Publishers.

Luckie, D. B., Aubry, J. R., Marengo, B. J., Rivkin, A. M., Foos, L. A., & Maleszewski, J. J. (2012). Less teaching, more learning: 10-yr study supports increasing student learning through less coverage and more inquiry. *Advances in Physiology Education, 36*, 325–335.

Madaus, J. (2000). Services for college and university students with disabilities: A historical perspective. *Journal of Postsecondary Education and Disability, 14*(1), 4–21.

McDaniel, N., Wolf, J., Mahaffy, C., & Teggins, J. (1994). Inclusion of students with disabilities in a chemistry laboratory course. *Journal on Post-Secondary and Education, 11*(1), 20–28.

McEwan, R. C., & Downie, R. (2013). College success of students with psychiatric disabilities: Barriers of access and distraction. *Journal of Postsecondary Education and Disability, 26*(3), 233–248.

McGuire, J. M., Scott, S. S., & Shaw, S. F. (2006). Universal design and its applications in educational environments. *Remedial and Special Education, 27*(3), 166–175.

MENTOR. (2009). *Elements of effective practice in mentoring* (3rd ed.). Retrieved from <http://www.mentoring.org/downloads/mentoring_1222.pdf>.

Metros, S., & C., Yang. (2006). *The importance of mentorship*. Cultivating Careers. EDUCAUSE.

Miner, D. L., Nieman, R., Swanson, A. B., & Woods, M. (Eds.), (2001). *Teaching chemistry to students with disabilities: A manual for high schools, colleges, and graduate programs.* (4th ed.). American Chemical Society. Retrieved from <http://www.acs.org/content/dam/acsorg/education/publications/teaching-chemistry-to-students-with-disabilities.pdf>.

Mohler, C. (2012). *The process of obtaining and retaining employment among the vision-restricted*. London: Western University. Available at: <http://ir.lib.uwo.ca/digitizedtheses>.

Moon, N. W., Todd, R. L., Morton, D., & Ivey, E. (2012). *Accommodating students with disabilities in science, technology, engineering, and mathematics (STEM): Findings from research and practice for middle grades through university education.* Atlanta: Center for Assistive Technology and Environmental Access, Georgia Institute for Technology.

Morgan, H., & Houghton, A. -M. (2011). *Inclusive curriculum design in higher education: Considerations for effective practice across and within subject areas*. The Higher Education Academy. Retrieved on Friday, March 27th, 2015 from <http://www.heacademy.ac.uk/assets/documents/inclusion/disability/ICD_introduction.pdf>.

Mullen, A. L., Goyette, K. A., & Soares, J. A. (2003). Who goes to graduate school? Social and academic correlates of educational continuation after college. *Sociology of Education*, *76*(2), 143–169.

Muller, L. (2006). Research collaboration with learning-disabled students. *Journal of College Science Teaching*, *36*(3), 26–29.

National Educational Association of Disabled Students. (2010). Success in STEM: Studying and pursuing a science or technology career as a post-secondary student with a disability.

National Science Foundation, National Center for Science and Engineering Statistics. (2013). Women, Minorities, and Persons with Disabilities in Science and Engineering: 2013. Special Report NSF 13-304. Arlington, VA.

Needs, S.E. (2002). *Special educational needs and disability act 2001*. Retrieved from <http://www.inclusivechoice.com/Special%20Educational%20Needs%20and%20Disability%20Act%202001.pdf>.

Neely, M. B. (2007). Using technology and other assistive strategies to aid students with disabilities in performing chemistry lab tasks. *Journal of Chemical Education*, *84*(10), 1697–1701.

Oakley, B., Parsons, J., & Wideman, M. (2012). *Identifying essential requirements: A guide for University Disability Professionals*. Kingston, Ontario: Queen's University.

Oberlander, S. E., & Spencer, R. J. (2006). Graduate students and the culture of authorship. *Ethics & Behavior*, *16*(3), 217–232.

Ontario Human Rights Commission. (n.d.) *Ontario Human Rights Code*. Retrieved from <https://www.ontario.ca/laws/statute/90h19>.

Ontario Human Rights Commission. (2003). The opportunity to succeed: Achieving barrier-free education for students with disabilities.

Ontario Human Rights Commission. (2004). Guidelines on accessible education.

Ontario Human Rights Commission. (2010). Ontario Human Rights Commission submission regarding the Ministry of Community and Social Services Proposed Integrated Accessibility Regulation under the Accessibility for Ontarians with Disabilities Act 2005.

Ontario Ministry of Community and Social Services. (2005). *Accessibility for Ontarians Disability Act*. Retrieved from <https://www.ontario.ca/laws/statute/05a11>.

Ontario Ministry of Education (OME). (2013). *Learning for all: A guide to effective assessment and instruction for all students, Kindergarten to Grade 12*. Retrieved from <http://www.edu.gov.on.ca/eng/general/elemsec/speced/LearningforAll2013.pdf>.

Opitz, D. L., & Block, L. S. (2006). Universal learning support design: Maximizing learning beyond the classroom. *Implementing Universal Design in Higher Education*, 205.

Pardo, P., & Tomlinson, D. (1999). *Implementing academic accommodations in field/practicum settings*. Calgary, Alberta: University of Calgary.

Pavri, S. (2010). *Effective assessment of students: Determining responsiveness to instruction*. New York, NY: Pearson.

Pence, L. E., Workman, H. J., & Riecke, P. (2003). Effective laboratory experiences for students with disabilities: The role of a student laboratory assistant. *Journal of Chemical Education*, *80*(3), 295–298.

Raynal, F., & Rieunier, A. (1998). *Pédagogie: dictionnaire des concepts clés: Apprentissage, formation, psychologie cognitive*. Paris: Édition Sociale Française (ESF).

Reeser, L. C. (1992). Students with disabilities in practicum: What is reasonable accommodation? *Journal of Social Work Education*, *28*(1), 98–109.

Rhodes, J. & DuBois, D.L. (2006) *Understanding and facilitating youth mentoring. Social Policy Report: Giving child and youth development knowledge away*. Retrieved from <http://www.srcd.org/sites/default/files/documents/20-3_youth_mentoring.pdf> (link is external) (PDF, 20 pages).

Roberts, B. (2012). Beyond psychometric evaluation of the student—task determinants of accommodation: Why students with learning disabilities may not need to be accommodated. *Canadian Journal of School Psychology, 27*(1), 72–81.

Roberts, B. (2013). *A lifeline for disability accommodation planning: How models of disability and human rights principles inform accommodation and accessibility planning.* Kingston: Queen's University. Available at <http://hdl.handle.net/1974/7806>.

Rose, D. H., Harbour, W. S., Johnston, C. S., Daley, S. G., & Abarbanell, L. (2006). Universal design for learning in postsecondary education: Reflections on principles and their application. *Journal of Postsecondary Education and Disability, 19*(2), 17.

Rose, D.H., & Meyer, A. (2002). Teaching every student in the digital age: Universal design for learning. Retrieved July 12, 2007, from the Association for Supervision and Curriculum Development Retrieved from: <http://www.cast.org/teachingev-erystudent/ideas/tes>.

Rose, M. (2009). *Accommodating graduate students with disabilities.* Toronto: Council of Ontario Universities.

Rose, R. R., & Fischer, K. (1998). Do authorship policies impact students' judgments of perceived wrongdoing? *Ethics & Behavior, 8,* 59–79.

Rowlett, E. J. (2008). Accessibility in MSOR: One student's personal experience. *MSOR Connections, 8*(1), 27–30.

Rowlett, E. J., & Rowlett, P. J. (2009). Visual impairment in MSOR. *MSOR Connections, 9*(4), 43–46.

Rowlett, P. J., & Rowlett, E. J. (2012). Experiences of students with visual impairments. In E. Cliffe & P. Rowlett (Eds.), *Good practice on inclusive curricula in the mathematical sciences.* (pp. 9–13). The Higher Education Academy.

Schelly, C. L., Davies, P. L., & Spooner, C. L. (2011). Student perceptions of faculty implementation of universal design for learning. *Journal of Postsecondary Education and Disability, 24*(1), 17–28.

Schleppenbach, D., & Lloyd, L. L. (2010). Increasing STEM accessibility in students with print disabilities through MathSpeak. *Journal of Science Education for Students with Disabilities, 14*(1), 1–15.

Sharby, N., & Roush, S. E. (2008). The application of universal instructional design in experiential education. In J. L. Higbee & E. Goff (Eds.), *Pedagogy and student services for institutional transformation: Implementing universal design in higher education* (Ch. 25, pp. 305–320). Minneapolis: University of Minnesota.

Sharby, N., & Roush, S. E. (2009). Analytical decision-making model for addressing the needs of allied health students with disabilities. *Journal of Allied Health, 38*(1), 54–62.

Shaw, R. (2011). Employing universal design for instruction. *New Directions for Student Services, 134,* 21–33.

Skarakis-Doyle, E., & McIntyre, G. (2008). *Western guide to graduate supervision.* London, ON: The University of Western Ontario Teaching Support Centre.

Sowers, J. A., & Smith, M. R. (2004). Nursing faculty members' perceptions, knowledge, and concerns about students with disabilities. *Journal of Nursing Education, 43*(5), 213–218.

Stodden, R. A., Conway, M. A., & Chang, K. B. T. (2003). Findings from the study of transition, technology and postsecondary supports for youth with disabilities: Implications for secondary school educators. *Journal of Special Education, 18*(4), 29–44.

Street, C. D., Koff, R., Fields, H., Kuehne, L., Handlin, L., GeMy, M., et al. (2012). Expanding access to STEM for at-risk learners: A new application of universal design for instruction. *Journal of Postsecondary Education and Disability, 25*(4), 363–375.

Sukhai, M.A., & Mohler, C.E., Doyle, T., Carson, E., et al. (2014a). *Creating an accessible science laboratory environment for students with disabilities.* Retrieved from <http://www.accessiblecampus.ca/wp-content/uploads/2014/06/Creating-an-Accessible-Science>.

Sukhai, M., Mohler, C., & Smith, F. (2014b). *Understanding accessibility in practical space learning environments across disabilities.* Council of Ontario Universities. Available online: <http://www.accessiblecampus.ca/>.

Sukhai, M.A., Duffett, E.M., Latour, A.R., Mohler, C.E., Pai, A., Gibson, D.N.E., et al. (2016) Deconstructing accessibility in graduate education for students with disabilities in Canada: Final report and recommendations of the National Graduate Experience Taskforce. Published July 2016: <http://neads.ca/en/about/projects/graduate-taskforce/index.php>.

Supalo, C. A., Mallouk, T. E., Rankel, L., Amorosi, C., & Graybill, C. M. (2008). Low-cost laboratory adaptations for precollege students who are blind or visually impaired. *Journal of Chemical Education, 85*(2), 243–247.

Tierny, J.P., Grossman, J.B., & Resch, N.L. (1995). *Making a difference: An impact study of Big Brothers Big Sisters.* Public/Private Venture. Retrieved from <http://www.ppv.org/ppv/publications/assets/111_publication.pdf>.

Tinto, V. (1993). *Leaving college: Rethinking the causes and cures of student attrition* (2nd ed.). Chicago,IL: The University of Chicago Press.

Tomlinson, C. A. (1999). *The differentiated classroom: Responding to the needs of all learners.* Alexandria, VA: ASCD.

Tsang, A. (2011). Students as evolving professionals: Turning the hidden curriculum around through the threshold concept pedagogy. *TD, 4*(3), 1–11. [Ref list].

Tsai, J. W., & Muindi, F. (2016). Towards sustaining a culture of mental health and wellness for trainees in the biosciences. *Nature Biotechnology, 34*(3), 353–355.

U.S. Department of Labor. Office of Disability Employment Policy. (n.d.). Cultivating leadership: Mentoring youth with disabilities. Retrieved from <http://www.dol.gov/odep/pubs/fact/cultivate.htm>.

United Nations Educational, Scientific and Cultural Organization. (2005). *Guidelines for inclusion: Ensuring access to education for all.* Paris, France: United Nations Educational, Scientific and Cultural Organization.

United Nations Convention on the rights of person with disabilities. (2006). Retrieved from <http://www.un.org/disabilities/convention/conventionfull.shtml>.

Weaver, G. C., Burgess, W. D., Childress, A. L., & Slakey, L. (Eds.), (2008). *Transforming institutions: Undergraduate STEM education for the 21st century.* West Lafayette, IN: Purdue University Press.

Wei, X., Yu, J. W., Shattuck, P., McCracken, M., & Blackorby, J. (2013). Science, technology, engineering, and mathematics (STEM) participation among college students with an autism spectrum disorder. *J. Autism Dev. Disord., 43*(7), 1539–1546.

Weiss, C. S. (1981). The development of professional role commitment among graduate students. *Human Relations, 34*(1), 13–14.

Wray, J., Fell, B., Stanley, N., Manthorpe, J., & Coyne, E. (2005). *The PEdDS project: disabled social work students and placements.* Hull: University of Hull.

Zhang, D., Landmark, L., Reber, A., Hsu, H., Kwok, O. M., & Benz, M. (2010). University faculty knowledge, beliefs, and practices in providing reasonable accommodations to students with disabilities. *Remedial and Special Education, 31*(4), 276–286.

Index

Note: Page number followed by "*f*" refer to figure.

A

ABIL. *See* Accessible Biomedical
    Immersion Laboratory (ABIL)
Academic
    academic integration, supervisors for
        students with, 172
    accommodations, 123
    environment, 97
    and professional standards maintenance,
        279–280
    rigor, 294–295
Access Scope, 235–236
Accessibility, 245–246
    in science laboratories, 261–262, 286
    solutions for accessibility concerns, 290
Accessibility for Ontarians with Disabilities
    Act (AODA), 267–268
Accessible Biomedical Immersion
    Laboratory (ABIL), 260–261
Accessible formats, 241
    accessibility and online learning
        environments, 245–246
    accessing in classroom setting, 242–243
    accessing in laboratory setting, 243–244
    availability of, 32
    Marrakesh Treaty, 246–247
    materials and technologies, 241–242
    print disabilities, 240
    reading print material, 240
    tips for accessing accessible formats in
        classroom and laboratory, 244
Accessible laboratory equipment, 259
Accessible learning environments, 286.
    See also Practical spaces; Science,
    technology, engineering, and
    mathematics (STEM)
    collaboration, 293–294
    dynamism, 293
    guiding principles for designing, 290–291

"Accessible safety tour" of lab, 261–262
Accommodation needs, 85, 87–88, 106
    disclosure, 74
    identification in nontraditional learning
        environments, 180–182
Accommodation(s), 6, 73, 96, 120. *See also*
    Student accommodations
    assessment/recommendations, 276
    conversations, 123
    in fieldwork setting, 203–204
    in graduate research laboratory, 202–203
    individualized, 145–148, 268, 293, 295
    interface with research integrity, 171–172
    legislation and duties to, 267–268
    mythbusting, 125–127
    partnerships for practicum placement
        learning, 272–274
        practicum site, 274
        student, 273
        university faculty, 273
        university services, 273–274
    personalized, 294
    plan, 63
    reasonable accommodation and undue
        hardship, 123–124
    relationship with essential requirements,
        122–123
    simulation learning as, 253–254
    stress, 99–100
    UDL and need for, 135–136
ACTion model, 107
ADAA. *See* Americans with Disabilities
    Amendments Act (ADAA)
Advocate, 86, 160
    student as, 87–88
Advocate, communicator, and trailblazer
    (ACTor), 194
AHRC. *See* Alberta Human Rights
    Commission (AHRC)

Alberta Human Rights Commission
    (AHRC), 122
Altered time, 281
Americans with Disabilities Act (1990),
    267–268
Americans with Disabilities Amendments
    Act (ADAA), 267–268
Anatomy teaching labs, microscope slide
    scanners in, 224–225
AODA. *See* Accessibility for Ontarians with
    Disabilities Act (AODA)
Archival spaces, 289. *See also* Accessible
    learning environments; Practical
    spaces
Assessment, 141
    of learning, 141
    planning, 141–142
Assisted learning, 250–251
Assistive technology (AT), 27, 31, 232–233,
    244. *See also* Mainstream technology
    barriers to accessing, 234–235
    challenges, 236
    in laboratory setting, 235–236
    low-tech AT, 233
Attitudes, 35–36
Audio format, 242
Authorship issues, 171–172

B

Barriers, 41, 240
    facing by students with disabilities, 27
    adapting mainstream technologies,
        31–32
    attitudes, 35–36
    availability of accessible formats, 32
    competing priorities in education,
        36–37
    lacking of access, 31
    "male-dominated" studies, 27
    occupational choice, 28
    self-advocacy, 34
    structural differences in student support
        systems, 29–30
    student awareness of support systems, 30
    student engagement with support
        systems, 30–31
    support network advocacy, 35
    universal accessing to scientific
        materials, 32

gatekeeper function, 37–38
    challenge of inductive reasoning, 38
    challenge of information lacking, 38
    challenge of misinformation, 37–38
    lacking of professional development
        for educators, 28–29
        for service providers, 29
    logistical considerations, 33–34
        in STEM classroom, 34
        in STEM laboratory, 33–34
Behavioral indicators, 101
Best practices, 260–261
BFAR. *See* Bona fide academic requirements
    (BFAR)
BFOR. *See* Bona fide occupational
    requirements (BFOR)
Biological psychology labs, 289–290
Biological sciences research lab, 224
Bona fide academic requirements (BFAR),
    120
Bona fide occupational requirements
    (BFOR), 17, 120
Boundary issues, 173
Braille, 54, 241

C

Calm and approachable mentors, 154
Canadian Charter of Rights and Freedoms,
    267–268
Canadian National Institute for the Blind
    (CNIB), 43–44
Captioning, 242
Career events, networking through, 192
Career mentorship, 191
CCTVs. *See* Closed circuit televisions
    (CCTVs)
Champion, 160
Checklist, 259
Classroom indicators, 101
Clinical education. *See* Practicum
    placements
Clinical lab learning environments, 289–290
Closed circuit televisions (CCTVs), 31, 225,
    233
CNIB. *See* Canadian National Institute for
    the Blind (CNIB)
Collaboration, 293–294
Communicator, 86
    student as, 88–89

Computer technology advancement, 134
Computer-aided instrumentation, 225
Computer-aided simulation, 250, 253
Conceptualizing inclusion, 141
Conferences, networking through, 192
Contingency assessment and planning, 204
Conventional wisdom, 10
"Cookie cutter" laboratories, 7
Coping strategies, 110
Course instructor, 267, 271, 276, 279–281
    initial meeting of, 274–275
Creative accommodation solutions, 146
Creative adaptations, 223
Creative mentors, 154, 159
Cross-disability environments, 109–110
CRPD. *See* United Nations Convention on
    Rights of Persons with Disabilities
    (CRPD)
Cultural perspectives about disability,
    20–21
Culture of accessibility, 11–12

D

Decision-making, 201
Descriptive video service (DVS), 242
Dialogue, 41
Differentiated instruction, 140, 144–145
Differentiated learning model, 145
Digital text, 134
Disabilities, Opportunities, Internetworking,
    and Technology program (DO-IT
    program), 13
Disability, 5–6, 85–86, 96–97, 166–167
    disability-related funding, 74
    disability-related needs, 200
    disability-specific networking, 192
    disability-specific structured professional
        networking activities, 191–192
    disclosure, 74
    faculty awareness of, 288–289
    management, 178
        stress, 99–100
    rights legislation, 19–20
    of STEM training, 100–101
    types, 18
Disability Discrimination Act (1995),
    267–268
Disability services office (DSO), 74
Discipline-specific requirements, 127

Disclosure, 73–74, 85, 124, 166–167
    accommodation need *vs.* disability, 74
    choice to disclose, 75–76
    impact of disability types, 79
    factors affecting, 80–81
    identifying right players, 78–79
    laboratory and practical space
        environments, 80
    process, 75
    pros and cons, 79–80
    rubric for, 78
    self-advocacy and, 76
    stress, 99–100
    students knowing priority, 77
    timing, 77–78
Distance learning. *See* E-learning
Diversity of practical spaces in STEM
    education, 289–290
DO-IT program. *See* Disabilities,
    Opportunities, Internetworking,
    and Technology program (DO-IT
    program)
DSO. *See* Disability services office (DSO)
DVS. *See* Descriptive video service (DVS)
Dynamism, 293

E

E-learning, 245–246
E-text, 241
Education, competing priorities in, 36–37
Education providers communication with
    students and service providers
    accounts of accommodation solutions, 59
    case studies
        graduate research environment ideal
            scenario, 65–66
        graduate research environment with
            pitfalls, 64–65
        undergraduate lab course-based ideal
            scenario, 63–64
        undergraduate lab course-based with
            pitfalls, 62–63
    disability services provider, 62
    faculty
        finding students with disabilities, 60
        members, 59
        responsibilities around communication
            and content delivery, 60–62
    teaching team, 62

Educators, 85, 96–97, 119–120
Effective accommodation principles
    individualization, 145
    individualized accommodations, 145–148
Electronic communications, 234
Emergency
    indicators, 102
    preparedness, 262
Employment for students in STEM graduate
        programs, 178–179
End-of-program licensing requirements, 127
Engineering custom technology solutions,
        228–229
Equality Act (2010), 267–268
Equipment, 200, 202, 204–205, 260–261
Equipment technology. See Off-the-shelf
        technology
Equitable opportunities for students, 281
Essential criteria, 200
Essential requirements, 62, 119, 210–212
    conversations, 123
    creative examination techniques, 121
    and evolution of sciences, 121–122
    gatekeeper function, 124–125
    measurement, 120
    mythbusting accommodations and,
        125–127
    relationship with accommodation, 122–123
    in STEM environments, 127–128
Evaluation, technical assistance, 219
Experiential education. See Practicum
        placements

F

Faculty, 77–78, 80, 87, 89, 96, 124–125,
        289, 294
    awareness of disability, 288–289
    members, 59, 86, 96, 101
    mentorship, 110, 157
FASEB. See Federation of American
        Societies of Experimental Biology
        (FASEB)
Federation of American Societies of
        Experimental Biology (FASEB), 128
Field-based learning. See Practicum
        placements
Fieldwork placements. See Practicum
        placements

Fieldwork setting, accommodation in, 203–204
Flexibility, 292–293
Flexible accommodations, 145–146
Food-science spaces, 289–290
Formal networking opportunities, 190–191
Frog's anatomy, 147
Funding issues, 173

G

Gatekeeper function, 21, 37–38, 124–125, 261
    challenge
        of inductive reasoning, 38
        of information lacking, 38
        of misinformation, 37–38
Geographical location, 281–282
Glass ceilings in STEM training pipeline,
        10–11
Graduate
    education, 81, 95
    employment environment, 178–179
    laboratory environments, 7–8
    research environment
        ideal scenario, 65–66
        with pitfalls, 64–65
Graduate research laboratory,
        accommodation in, 202–203
Group work, 143–144

H

Hidden curriculum, 142–144, 178
High-pressure environment, 95
High-tech AT, 233–234
Higher education in global landscape of
        disability rights legislation, 19–20
Histology teaching labs, microscope slide
        scanners in, 224–225
Homogenous practice area, 282
Human accommodation
    essential requirements, 210–212
    human assistant, 209–210
Human assistant, 209–210. See also
        Technical assistance
Human-rights framework, 17–18
IDP. See Independent development plan (IDP)

I

Implicit expectation identification, 142–144
"Impostor" syndrome, 95–99

In-person mentorship, 157
Inclusive teaching, 135. *See also* Universal design for learning (UDL)
  practices, 135, 140
    approaches to developing, 144–145
    identifying implicit expectations, 142–144
    principles of effective accommodation, 145–148
    principles of inclusion, 140–142
Independent development plan (IDP), 182–183
Individualized accommodations, 145–148, 268, 293, 295
Informal networking opportunities, 190–191
Informational barrier, 32
Informational interviews, 193
Institute for Accessible Science, 260–261
Instructional design, 140, 145
Integrated professional development opportunities, 190
Integrated Science laboratory (iScience laboratory), 260
Intellectual property legislation (IP legislation), 246
International practicums, 282–283
Internships. *See* Practicum placements
Interviews, 218
IP legislation. *See* Intellectual property legislation (IP legislation)
iPads, 226
iPhones, 226
iScience laboratory. *See* Integrated Science laboratory (iScience laboratory)

J

Job posting, 218

K

Kindergarten to grade 12 education system (K-12 education system), 29–30
Knowledge base, 228–229

L

Lab indicators, 101
Lab spaces, 258–259
Lab-based activities, 202
Laboratory setting, AT in, 235–236
Large-print materials, 241
Learning environments, 236

Learning management system (LMS), 59, 245
Learning outcomes, 121, 128
Light microscopy (LM), 235–236
Life stressors, 95
Lived experience, 41
LM. *See* Light microscopy (LM)
LMS. *See* Learning management system (LMS)
Logistical considerations, 33–34
  in STEM classroom, 34
  in STEM laboratory, 33–34
Long-distance peer support. *See* Peer-support networks, Virtual peer support
Low expectations, 97
Low-tech AT, 233
Low-tech solutions, 226–227

M

Mainstream technology, 223, 250–251. *See also* Assistive technology (AT)
  for accessibility, 226
  accessible mainstream technology solutions, 226
  adaptation, 31–32
  CCTV, 225
  determining best technology solution, 227
  engineering custom technology solutions, 228–229
  low-tech solutions, 226–227
  microscope slide scanners, 224–225
  robotics, 224
  student considerations, 228
  technological solutions, 223–224
  universal design in lab, 225
Marrakesh Treaty, 246–247
MathML, 245
Mechanical assistance, 250–253
Mental health and well-being for students, 96
  disability of STEM training, 100–101
  impostor syndrome, 97–99
  improving well-being and maintain balance, 102–103
  stress
    disclosure, accommodation, and disability management, 99–100
    recognizing signs, 101–102
    of STEM training, 100–101
    of trailblazer, 96–97

Mentee, 155
Mentor(s), 98, 110, 124, 156
  benefits, 160
  qualities, 158–160
    creative, 159
    open-minded, 159
    proactive, 158
    responsive, 159
  selection, 160–161
Mentorship, 159
  faculty, 153
  mentor qualities, 154
  formal relationship, 155
  forms, 156–158
    faculty mentorship, 157
    in-person mentorship, 157
    online mentorship, 157
    peer mentorship, 158
    senior mentorship, 157
  importance in STEM, 154–155
  relationship between two individuals, 155
  relationship forming between two people,
    155
  student interests, 153
Microscope slide scanners, 224–225
Mythbusting accommodations, 125–127

N

National context, 270
National Postdoctoral Association (NPA),
    128
National practicums, 282–283
National Science Foundation (NSF), 27
Necessary competencies. See Essential
    requirements
Networking, 190
  creating own portfolio, 193–194
  framing disability in, 194
  personal story or brand and impact on,
    194–195
  through conferences, career events, and
    symposia, 192
  types of opportunities, 190–191
Nontraditional learning environments, 178
  considerations, 180
  identifying accommodation needs in,
    180–182
  stress and, 183

NPA. See National Postdoctoral Association
    (NPA)
NSF. See National Science Foundation (NSF)

O

Occupational choice, 28
Occupational therapy, practical spaces in,
    287–289
OCR software. See Optical character
    recognition software (OCR software)
Off-the-shelf technologies, 31–32, 223
OHRC. See Ontario Human Rights
    Commission (OHRC)
One-on-one mentorship. See In-person
    mentorship
Online course platforms, 240, 245–246
Online learning. See E-learning
Online mentorship, 157
Ontario Human Rights Code, 267–268
Ontario Human Rights Commission
    (OHRC), 122
Open-minded faculty members, 154
Open-minded mentor, 159
Optical character recognition software (OCR
    software), 31, 234

P

PAPM. See Practicum Accommodations
    Process Model (PAPM)
Peer mentorship, 107–108, 110–111, 158
Peer support, 102
Peer-support networks, 106
  beginning of conversation, 111
  benefits, 109
  community development, 111
  operational issues, 110
  outside disability, 112–113
  peer mentorship vs., 110–111
  potential challenges, 109–110
  recipe, 107f
  student's mental and social health,
    106–107
  types, 108
  virtual peer support, 112
Peer networks and collaborations, 192–193
Personalized accommodations, 294
Physical access in science laboratories,
    33, 258

accessibility and safety, 261–262
best practices, 260–261
considerations for, 259
universal design and, 259–260
Physical indicators, 101
Physical therapy, practical spaces in,
    287–289
Physics laboratory, computer-aided
    instrumentation in, 225
Physiology teaching labs, microscope slide
    scanners in, 224–225
"Pioneer", 89
Positive relationships, 294
Postsecondary institution, 288–289
Power dynamic between student and
    supervisor, 169
Practical spaces, 287. *See also* Accessible
    learning environments; Archival
    spaces
differences with traditional science
    laboratories, 287
diversity in STEM education, 289–290
environments, 12
in occupational and physical therapy,
    287–289
Practicum Accommodations Process Model
    (PAPM), 268–270, 274–278. *See also*
    Practicum placements
accommodations assessment/
    recommendations, 276
considerations and strategies for
    successful provision, 278–281
    equitable opportunities for students, 281
    maintenance of academic and
        professional standards, 279–280
    time, 280–281
contextual considerations, 270–272
    national context, 270
    practicum site context, 271–272
    student context, 272
    university context, 270–271
exploring potential practicum partners,
    276–277
finalizing specific practicum
    arrangements, 277
initial meeting of student and course
    instructor, 274–275
monitoring performance throughout
    practicum, 277

partnerships for practicum placement
    learning, 272–274
    practicum site, 274
    student, 273
    university faculty, 273
    university services, 273–274
reflection on teaching and learning, 278
Practicum placements, 266–267. *See also*
    Practicum Accommodations Process
    Model (PAPM)
considerations for national and
    international practicums, 282–283
student accommodations
    altered time, 281
    geographical location, 281–282
    homogenous practice area, 282
    provision of one primary preceptor, 282
    use of technology, 282
Practicum site, 267, 274
    context, 271–272
Preceptors, 267, 271, 276–278
    provision of one primary, 282
Principles
    for active learning, 135
    of inclusion, 140–144
Print disabilities, 32
Proactive faculty members, 154
Proactive mentor, 158
Professional competencies, 267, 275, 279,
    294–295
Professional development, 192
    disability-specific structured professional
        networking activities, 191–192
    for educators, 28–29
    informational interviews, 193
    networking, 190
        creating own portfolio, 193–194
        framing disability in, 194
        personal story or brand and impact on,
            194–195
        through conferences, career events, and
            symposia, 192
        types of opportunities, 190–191
    peer networks and collaborations,
        192–193
    receptivity of network, 195–196
    for service providers, 29
    student, 190
"Professional disability pride", 111

"Professional" graduate programs, 81
Psychological syndrome, 97

R

Reasonable accommodation, 6
Rehabilitation Act, 267–268
Research assistantships, 8
Research integrity
    interface with accommodations, 171–172
"Research-stream" graduate programs, 81
Resource strengths, 272
Responsive mentors, 154, 159
"Right approach", 251
Robotics, 224

S

Safety
    indicators, 101–102
    requirements, 200
    in science laboratories, 261–262
Science, technology, engineering, and
        mathematics (STEM), 4–5, 13–14,
        17, 27, 59, 96, 106, 120, 154,
        165, 177, 234, 258, 286. *See also*
        Accessible learning environments
    academic rigor and professional
        competence, 294–295
    and disability, 297
    disability-specific accommodations, 160
    diversity of practical spaces in, 289–290
    education, 17–18
        CRPD, 18–19
        cultural perspectives about disability,
            20–21
        form of disability, 16–17
        higher education in global landscape of
            disability rights legislation, 19–20
        science as international endeavor, 21–22
    employment for students in graduate
        programs, 178–179
    encompasses many faces of student in,
        295
    environments, essential requirements in,
        127–128
    facing persons, 99
    fields, 86, 89, 97
    flexibility, 292–293
    fostering positive relationships, 294
    glass ceilings in STEM training pipeline,
        10–11
    guiding principles for designing
        accessible learning environments,
        290–291
    materials, 243
    mentorship importance in, 154–155
    online resources, 161
    programming, 140
    programs, 145–146
    STEM-based programs, 80–81
    students and trainees in, 178
    training, 95, 99–100
        disability, 100–101
        stresses, 100–101
Science as international endeavor, 21–22
Scientific materials universal accessing to,
        32
Scientific training, 17
Screenreaders software, 241–242
Self-advocacy, 34, 76, 87
    self-advocacy-based model, 99
Self-advocate, 48
Senior mentorship, 157
Service learning. *See* Practicum placements
Simulation learning, 250
    and accessibility, 253
    as accommodation, 253–254
    applying simulation learning to sciences,
        252–253
    as course/program component, 254
    and learning styles, 252
    in postsecondary education, 254
    scenarios, 251
    virtual learning in sciences, 251–252
Six-step collaborative decision-making
        model, 268
Smart Pen, 134
Smartphones, 223
Social integration, supervisors for students
        with, 172
Social network, 108
Soft skills, 190
Software solutions, 234
Special Education Needs and Disability Act
        (2001), 267–268
STEM. *See* Science, technology,
        engineering, and mathematics
        (STEM)

Stress
  of accommodation, 99–100
  of disability management, 99–100
  of disclosure, 99–100
  recognizing signs, 101–102
  of STEM training, 100–101
  of trailblazer, 96–97
Stressors, 95
Structured networking, 191–192
Student accommodations. *See also*
      Accommodations
  altered time, 281
  geographical location, 281–282
  homogenous practice area, 282
  provision of one primary preceptor, 282
  use of technology, 282
Student in leadership, mentorship, and
      supervision role
  achievement of necessary competencies,
      182–183
  disability management, 178
  employment for students in STEM
      graduate programs, 178–179
  identifying accommodation needs,
      180–182
  nontraditional learning environments, 178
  stress and nontraditional learning
      environments, 183
  student as trailblazer, 183–184
  student not in traditional learning
      environment, 179–180
  student's living experience with disability,
      184
Student perspectives on disability
  barriers in selecting subjects, 53–54
  career selection, 48–49
  concerning students opinion, 49
  encouragement, 45–47
  experience in schools and colleges, 47–52
  facilities for disable students, 42–43
  ignorance, 43–45
  knowledge, 52–53
  positive attitude, 53
Student support
  community, 272
  systems
    structural differences in, 29–30
    student awareness, 30
    student engagement with, 30–31

Student(s), 96, 273
  as ACTor, 87, 106
    secondary school to postsecondary
        transition, 90
    student as advocate, 87–88
    student as communicator, 88–89
    student as trailblazer, 89–90
  considerations, 228
  context, 272
  in crisis, 169–170
  disabilities, 95–96, 166, 169,
      261–262
  equitable opportunities for, 281
  initial meeting of, 274–275
  living experience with disability, 184
  management strategies, 99–100
  student-driven learning process, 81
  as trailblazer, 183–184
Students with disabilities
  accessibility and STEM, 13–14
  accommodation
    in fieldwork setting, 203–204
    in graduate research laboratory,
        202–203
    in science lab or fieldwork settings,
        204–205
  application of best practices across
      disciplines, 12–13
  CRPD, 4
  culture of accessibility, 11–12
  exclusive education, 6–7
  filling knowledge gap, 5
  glass ceilings in STEM training pipeline,
      10–11
  inclusive education, 6–7
  integrated education, 6–7
  lab-based activities, 202
  partnership among faculty, 204–205
  participation, 8–10
  practical space environments, 12
  reasonable accommodation, 6
  in sciences, 201
  segregated education, 6–7
  STEM fields, 5
  teaching practices, supports, and
      accommodations, 201
  undergraduate *vs.* graduate laboratory
      environments, 7–8
  work, 8

Summative assessment, 141
Supervision and mentorship roles, 165
Supervision relationship
  authorship issues, 171–172
  boundary issues, 173
  delegated supervision, 174
  disability, 166–167
  disclosure, 166–167
  funding issues, 173
  students in crisis, 169–170
  student–supervisor relationship, 166–167
    clarifying expectations in, 168–169
    deterioration in, 168
    foundation, 165–166
    quality, 167–168
  supervisor's knowledge
    in disability-related processes, 170
    interface between essential requirements
      and academic accommodations, 171
Supervisors, 99
  assisting students with academic and
    social integration, 172
  knowledge
    in disability-related processes, 170
    interface between essential requirements
      and academic accommodations, 171
"Supply side" approaches, 10–11
Support network advocacy, 35
Symposia, networking through, 192

T

Tactile graphics, 242
TAs. See Teaching assistantships (TAs)
Teaching assistants. See Teaching
    assistantships (TAs)
Teaching assistantships (TAs), 8, 28, 60–61,
    74, 202
Teaching lab setting, activities in, 202
Teaching practices, supports, and
    accommodations, 201
Teaching team, 62
Teaching tools, 251
Team science, 192–193
Technical assistance, 210, 212–213, 252–253
  credit sharing, 214
  defining role, 217–218
  evaluation, 219
  expensiveness, 215–216

interviews, 218
job posting, 218
student gaining appropriate learning from,
    216
student helping, 212–213
training, 218–219
unfair advantage, 212
unrealistic, 213–214
utilization, 216–217
Technically necessary requirements, 127
Technology use in student accommodations,
    282
"Thought rubric", 290
Time in accommodations process, 280–281
Traditional learning environment, 179–180.
    See also Nontraditional learning
    environments
Traditional science laboratories
  practical spaces differences with, 287
Trailblazers, 86
  stress of, 96–97
  student as, 89–90
Trainee, 171
Training, 218–219
  environments, 287
Transition points, 98

U

UD. See Universal design (UD)
UDI. See Universal design of instruction
    (UDI)
UDL. See Universal design for learning
    (UDL)
Undergraduate lab
  course-based ideal scenario, 63–64
  course-based with pitfalls, 62–63
Undergraduate laboratory environments, 7–8
Undergraduate STEM training, 100
Undue hardship, 6
United Nations Convention on Rights of
    Persons with Disabilities (CRPD), 4,
    18–19
Universal design (UD), 130, 132, 286
  approaches to, 135
  in classroom, 136
  computer-aided instrumentation, 225
  as conceptual and philosophical
    foundation, 133

goal, 133
laboratories, 136
and physical accessibility of science
    laboratories, 259–260
principle, 291–292
proactive approach, 136
Universal design for learning (UDL), 130,
    144, 291–292
applying to learning environment,
    133–134
barriers to learning, 131–132
components, 130
identifying principles for planning
    instruction, 144–145
and need for accommodation,
    135–136
principles, 131–133
technology in implementation, 134
Universal design of instruction (UDI),
    279–280, 291–292

Universal instructional design (UID). *See*
    Universal design of instruction (UDI)
Universally designed environment, 295
University
    context, 270–271
    course calendars, 271
    faculty, 273
    services, 273–274

V

Virtual learning, 250–252

W

Web-based training. *See* E-learning
Webmasters, 32
Windowing, 242
World Health Organization, 205
World Intellectual Property Organization
    (WIPO), 246